Lecture Notes in Mathematics

Edited by A. Dold and B. Eckmann

1230

Numerical Analysis

Proceedings of the Fourth IIMAS Workshop held at Guanajuato, Mexico, July 23–27, 1984

Edited by J. P. Hennart

Springer-Verlag
Berlin Heidelberg New York London Paris Tokyo

Editor

Jean-Pierre Hennart
IIMAS-UNAM
Apartado Postal 20-726
01000 México, D.F., México

Mathematics Subject Classification (1980): 65F; 65K; 65L; 65M; 65N;

ISBN 3-540-17200-9 Springer-Verlag Berlin Heidelberg New York
ISBN 0-387-17200-9 Springer-Verlag New York Berlin Heidelberg

Printing and binding: Druckhaus Beltz, Hemsbach/Bergstr.
2146/3140-543210

To the memory of our colleague, David Alcaraz,

who died in Mexico's earthquake,

on September 19th, 1985.

FOREWORD

During the five days 23rd-27th July 1984 in Guanajuato, Guanajuato, México, the Institute for Research in Applied Mathematics and Systems (IIMAS) of the National University of Mexico (UNAM) held its Fourth Workshop on Numerical Analysis. As in the first three versions in 1978, 1979 and 1981, the program of this research workshop concentrated on the numerical aspects of three main areas, namely optimization, linear algebra and differential equations, both ordinary and partial. J.H. Bramble, J.R. Cash, T.F. Chan, J.E. Dennis, Jr., J. Douglas, Jr., H.C. Elman, R. England, R.S. Falk, D. Goldfarb, A. Griewank, S.P. Han, J.P. Hennart, A.V. Levy, R.D. Skeel, M.F. Wheeler and M.H. Wright were invited to presente lectures. In total 29 papers were delivered, of which 18 are offered in these Proceedings.

Like the Third Workshop, this one was supported by a generous grant from the Mexican National Council for Science and Technology (CONACyT) and the U.S. National Science Foundation, and was part of the Joint Scientific and Technical Cooperation Program existing between these two countries. In relation to this essential funding aspect, it is a pleasure to express again my thanks to R. Tapia, of the Mathematical Sciences Department at Rice, for his continual advice and help prior to the workshop. This time in particular, as the confirmation of the funding was very close to the beginning of the workshop, his role was fundamental in providing us with the above excellent list of invited speakers from the U.S.

My thanks also go to S. Gómez of IIMAS for the enthusiasm and energy she displayed at the local arrangements level, to my colleagues of the Numerical Analysis Department for their friendly cooperation and to IIMAS for its continuous support. Finally, I would like to acknowledge the invaluable help of Ms. A. Figueroa in the typing and retyping needed to transform a set of manuscripts into book form.

Mexico City, November 1985

J.P. HENNART

CONTENTS

LIST OF PARTICIPANTS

BOGGS, P.T.	Center for Applied Mathematics, National Bureau of Standards, Gaithersburg, Maryland 20899, USA.
BRAMBLE, J.H.	Department of Mathematics, Cornell University, Ithaca, New York 14853, USA.
BRENIER, Y.	INRIA Rocquencourt, 78150 Le Chesnay, France.
BREZZI, F.	Dipartimento di Meccanica Strutturale, University of Pavia and Istituto di Analisi Numerica, C.N.R., 27100 Pavia, Italy.
BYRD, R.	Department of Computer Science, University of Colorado, Boulder, Colorado 80309, USA.
CALDERON, A.	IIMAS-UNAM, Apdo. Postal 20-726, 01000 México, D.F., Mexico.
CASH, J.R.	Department of Mathematics, Imperial College, London SW7 2BY, England.
CHAN, T.F.	Department of Computer Science, Yale University, New Haven, Connecticut 06520, USA.
CHAVENT, G.	INRIA Rocquencourt, 78150 Le Chesnay, France.
CHEN, B.	IIMAS-UNAM, Apdo. Postal 20-726, 01000 México, D.F., Mexico.
DENNIS, Jr., J.E.	Mathematical Sciences Department, Rice University, Houston, Texas 77001, USA.
DOUGLAS, Jr., J.	Department of Mathematics, The University of Chicago, Illinois 60637, USA.
ELMAN, H.C.	Department of Computer Science, Yale University, New Haven, Connecticut 06520, USA.
ENGLAND, R.	IIMAS-UNAM, Apdo. Postal 20-726, 01000 México, D.F., Mexico.
FALK, R.S.	Department of Mathematics, Rutgers University, New Brunswick, New Jersey 08903, USA.
GAY, D.M.	Bell Laboratories, Murray Hill, New Jersey 07974, USA.
GOLDFARB, D.	Department of Industrial Engineering and Operations Research, Columbia University, New York, New York 10027, USA.
GRIEWANK, A.	Southern Methodist Unversity, Dallas, Texas 75275, USA.
GOMEZ, S.	IIMAS-UNAM, Apdo. Postal 20-726, 01000 México, D.F., Mexico.
HAN, S.P.	Mathematics Department, University of Illinois Urbana, Illinois 61801, USA.
HENNART, J.P.	IIMAS-UNAM, Apdo. Postal 20-726, 01000 México, D.F., Mexico.
LEVY, A.V.	IIMAS-UNAM, Apdo. Postal 20-716, 01000 Méxi o, D.F., Mexico.

MARINI, L.D.	Istituto di Analisi Numerica, C.N.R., 27100 Pavia, Italy.
MORALES, J.L.	IIMAS-UNAM, Apdo. Postal 20-726, 01000 México, D.F., Mexico.
NOCEDAL, J.	Department of Electrical Engineering and Computer Science, Northwestern University, Evanston, Illinois 60201, USA.
SARGENT, R.W.H.	Chemical Engineering Department, Imperial College, London SW7 2BY, England.
SKEEL, R.D.	Department of Computer Science, University of Illinois, Urbana, Illinois 61801, USA.
TAPIA, R.	Department of Mathematical Sciences, Rice University, Houston, Texas 77001, USA.
VARGAS, C.	Departamento de Matemáticas, Centro de Investigación y de Estudios Avanzados del IPN, Apdo. Postal 14-740, 07000 México, D.F., Mexico.
WATSON, L.T.	Department of Computer Science, Virginia Polytechnic Institute and State University, Blacksburg Virginia 24061, USA.
WHEELER, M.F.	Department of Mathematical Sciences, Rice University, Houston, Texas 77001, USA.
WRIGHT, M.H.	Department of Operations Research, Stanford University, Stanford, California 94305, USA.

A GLOBAL ZERO RESIDUAL LEAST SQUARES METHOD

S. GOMEZ, A.V. LEVY and A. CALDERON
IIMAS-UNAM, Numerical Analysis Dept.
Apartado Postal 20-726
01000 México, D.F.
MEXICO

INTRODUCTION

In this work we want to find the least squares solution of a system of nonlinear equations

$$f_i(x)=0 \qquad i=1,\dots,m$$

where $x \in \mathbb{R}^n$, $f_i:\mathbb{R}^n \to \mathbb{R}$ and $m \geq n$. To solve this problem we seek for a minimum of the function $F(x)$, that is

$$\min_x F(x)=f^T(x)\ f(x)$$

In general there will exist local minima x^* of this function with small residuals ($F(x^*) \neq 0$), but in this paper we shall assume that the zero residual solution ($F(x^*)=0$) also exists. It is this global solution the one that is of interest in the present work and will be referred as the global least squares solution. In order to avoid all local minima of $F(x)$ we shall use a deflation technique called the tunneling function which preserves the global solution of $F(x)$. In order to find this solution the Gauss-Newton Method will be used.

The present method is not only able to avoid local solutions but also has the nice property of handling rank one defiencies of the Jacobian $J(x)$ of $f(x)$, which is a typical difficulty for the Gauss-Newton Method.

1. STATEMENT OF THE PROBLEM

We want to minimize a sum of squares

$$\min_x F(x)=f^T(x)\ f(x) \tag{1.1}$$

If $J(x)$ is the Jacobian of $f(x)$, then the gradient of $F(x)$ will be

$$g(x) = 2J^T(x)\ f(x) \tag{1.2}$$

Problem (1.1) has a local solution at x^* if

$$J^T(x^*)\ f(x^*) = 0 \tag{1.3}$$

and it is the global zero residual solution if

$$F(x^*_G) = f^T(x^*_G)\ f(x^*_G) = 0 \tag{1.4}$$

If $G_i(x)$ is the Hessian of $f_i(x)$, then the Hessian of $F(x)$ will be

$$G(x) = 2J^T(x)\ J(x)+2 \sum_{i=1}^{m} f_i(x)\ G_i(x) \tag{1.5}$$

In practice, for small residual and for zero residual problems

$$G(x) \simeq 2J^T(x)\ J(x) \tag{1.6}$$

It is this approximation the one we shall use in the present work.

2. THE TUNNELING FUNCTION CONCEPT

In order to avoid the local solutions of problem (1.1), we will now solve

$$\min_x \phi(x) = T^T(x)T(x) \tag{2.1}$$

where $T(x)$ is the tunneling function defined as

$$T(x) = \frac{f(x)}{[(x-x^p)^T(x-x^p)]^k} = \frac{f(x)}{(\|x-x^p\|_2^2)^k} \tag{2.2}$$

and its Jacobian

$$T_x(x) = \frac{1}{(\|x-x^p\|_2^2)^k}\left[J(x) - \frac{2k}{(\|x-x^p\|_2^2)} \quad f(x)(x-x^p)^T\right] \tag{2.3}$$

Obviously if the parameter k is zero then $T(x)\equiv f(x)$ and $T_x(x)\equiv J(x)$. Also it is clear that, $T_x(x)$ is the Jacobian of the original function $J(x)$ plus a rank one matrix. From the definition (2.2) it is very easy to show that the global solution for $\phi(x)$, $\phi(x^*_G)=0$ is the global solution for the original problem $F(x^*_G)=0$.

3. FEATURES OF THE ALGORITHM

3.1 The Gauss-Newton Step

The Gauss-Newton step for solving problem (1.1) will be

$$J(x)^T J(x)p = -J(x)^T f(x) \quad , \tag{3.1}$$

and in order to avoid ill conditioning of $J(x)^T J(x)$, a better definition of p is obtained by finding the least squares solution of

$$J(x)p = -f(x) \tag{3.2}$$

where p minimizes $\|J\,p+f\|_2$.

The same consideration applies for solving problem (2.1), getting the Gauss-Newton step for the tunneling function

$$T_x(x)\,p = -T(x). \tag{3.3}$$

In order to obtain the solution of the systems (3.2) and (3.3) we can use the singular value decomposition of J and T:

$$J = U \begin{bmatrix} S \\ \hline 0 \end{bmatrix} V^T \quad , \quad \text{if} \quad k=0$$

or

$$T = U \begin{bmatrix} S \\ \hline 0 \end{bmatrix} V^T \quad , \quad \text{if} \quad k \neq 0$$

where $S=\mathrm{diag}(\sigma_1, \sigma_2, \ldots, \sigma_n)$ is the matrix of singular values with $\sigma_i \geq 0$, U is an m×m orthogonal matrix and V is an n×n orthogonal matrix. Then the least squares solution for systems (3.2) and (3.3) are given by

$$p = -VS^{-1}V^T f \quad , \quad \text{if } k = 0$$

or

$$p = -VS^{-1}V^T T \quad , \quad \text{if } k \neq 0$$

where

$$S^{-1} = \begin{cases} \dfrac{1}{\sigma_j} & \text{if} \quad \sigma_j \neq 0 \\ 0 & \text{if} \quad \sigma_j = 0. \end{cases}$$

3.2 Parameter Computation

We start the algorithm using the step for the original system defined in Eq. (3.2), until we reach a point say x^P, at which one detects either a rank defficiency of the Jacobian $J(x^P)$ (singular point) or a local solution $J^T(x^P)f(x^P)=0$ (critical point). At this point x^P, we deflate the original system using the tunneling function Eq. (2.2) (in practice this means that k will take a value different from zero, creating a pole at x^P with strength k). We then proceed using the step defined in Eq. (3.3).

Starting with k=0.1 and increasing k with $\Delta k=0.1$, the algorithm computes the appropriate non zero value of k to get a descent Gauss-Newton step for T(x). To avoid division by zero when using Eq. (2.2) and (2.3), the above Gauss-Newton step is computed at the point

$$x = x^p + \varepsilon r \tag{3.4}$$

where ε is a very small parameter so that x is in a neighborhood of x^p, and r is a random vector $r_i \in [-1,1]$. Good results are obtained if $\varepsilon=0.1$. See Ref. [1] for detailed description of the parameters.

Once k is non zero, according to Eq. (2.2) if $\|x-x^p\|$ becomes larger than one, the shape of T(x) becomes very flat, slowing convergence. Therefore, if at some iterand x, the distance $\|x-x^p\|_2 \geq 1$, we move the position of the pole x^p along the vector $(x-x^p)$ so that $\|x-x^p\|=\varepsilon$. In this fashion we shall always have

$$\|x - x^p\|_2 \leq 1 \quad . \tag{3.5}$$

In the other hand, having $\|x-x^p\|_2 \leq 1$ and $k \neq 0$ leads to a situation where $\|T(x)\|_2^2 > \|f(x)\|_2^2 = F(x)$. Therefore, in order to improve convergence, by reducing T(x) it is desirable to reset k to zero as soon as possible and then proceed on the original system using the step defined in Eq. (3.2). This can be done whenever the norm of the residuals $F=\|f(x)\|_2^2$ drops below the level of the norm at the point where k was increased from zero.

3.3 Main Features of the Algorithm

We want to point out here the main features of the algorithm which are:

a) It can handle rank-one defficiency of the Jacobian (singular points).

b) It does not stop at local solutions, and proceeds until it gets the global solution.

Briefly let us see how the tunneling idea achieves these features:

a) At singular points where the Jacobian has a rank-one deficiency, the solution $J(x)p=-f(x)$ is not unique (p is arbitrarily large), but if we choose x, so that $(x-x^p)$ is not orthogonal to the null space of J(x), then $T_x(x)$ has full rank and $T_x(x)p=-T(x)$ can be solved.

b) At critical points $J^T(x)f(x)=0$, the Gauss Newton step of Eq. (3.1) is not defined. However, as stated in section 3.2 when this occurs k takes a value different from zero; then the expression

$$T_x^T(x)T(x)= \frac{1}{(\|x-x^p\|_2^2)^{2k}}\left[J^T(x)f(x)- \frac{2k}{\|x-x^p\|_2^2}(x-x^p)f^T(x)f(x)\right] \quad (3.6)$$

shows that $T_x^T(x)T(x) \neq 0$ unless x is the global solution, that is $f^T(x)f(x)=0$, and therefore the Gauss-Newton step Eq. (3.3) is well defined, and since it is a descent direction for problem (2.1), the algorithm proceeds to the global solution.

There is another feature of the algorithm which is worth mentioning: when k is different from zero (because of a detection of a singular or a critical point somewhere before), the algorithm does not necessarily detect at x if $J^T(x)f(x)=0$, because $T_x^T(x)T(x) \neq 0$. This fact is important because the method approaches the global solution without the need to locate local solutions as was the case in our previous work in global optimization, Ref. [2], [3] and [4].

The value of k is calculated to get a descent Gauss-Newton step for system (2.2), but if k is not sufficiently large, one could also reach a critical point of the system T(x), that is

$$T_x(x)^T T(x) = 0$$

which is not a critical point of system f(x), that is $J(x)^T f(x) \neq 0$.

However, from Eq. (3.6) it can be seen that increasing k will be enough to get

$$T_x^T(x)T(x) \neq 0.$$

Geometrically it means that for k sufficiently large the error function $\phi(x)=T^T(x)T(x)$ stretches out.

4. Numerical Examples

Several numerical examples were solved, in order to test the method, which are reported in Ref. [1]. In this paper we illustrate only one of those examples.

Consider the problem (Cragg-Levy)

$$\begin{aligned} f_1 &= (e^{x_1}-x_2)^2 \\ f_2 &= 10(x_2-x_3)^3 \\ f_3 &= [\operatorname{sen}(x_3-x_4)/\cos(x_3-x_4)]^2 \\ f_4 &= x_1^4 \\ f_5 &= x_4-1 \end{aligned} \tag{4.1}$$

for which we have found the following local minima

$$\begin{aligned} &x=(0,1,0,0) && \text{with } f^T(x)f(x)=1.01\times10^2 \\ &x=(0,1,1,0) && \quad\quad . \quad =6.8\times10^0 \\ &x=(-0.631,0,0,0) && \quad\quad . \quad =1.1\times10^0 \\ &x=(0.552,2,2,2) && \quad\quad . \quad =1.01\times10^0 \\ &x=(-1.110,-2,-2,-2) && \quad\quad . \quad =4.07\times10^1 \end{aligned} \tag{4.2}$$

and the global minimum

$$f^T(x)f(x) = 0 \qquad \text{at} \qquad x=(0,1,1,1) \tag{4.3}$$

The above local minima were found using a Levenberg-Marquard algorithm (Moré's version) when the following initial points were used

$$\begin{aligned} x_i = &(0,0,0,0),\ (0,1,0,0,),\ (0,1,1,0) \\ &(1,2,2,2),\ (-1,-2,-2,-2) \end{aligned} \tag{4.4}$$

Obviously at the local minima the Levenberg-Marquard algorithm terminated since it is a local method.

The tunneling algorithm starting from the same initial points (4.4) arrived at the global solution requiring the computing effort given in the following table:

Initial Point	Iter	fn	Jac	
(0, 0, 0, 0)	6	9	8	
(0, 1, 0, 0)	8	11	10	
(0, 1, 1, 0)	7	10	9	Final error 10^{-6}
(1, 2, 2, 2)	8	13	10	
(-1,-2,-2,-2)	8	12	10	

TABLE I. Numerical results for example (4.1), showing the required number of iterations, function evaluations and Jacobian evaluations for the present method to reach the global zero residual solution.

On its way to the global minimum the present method detected the following points as "singular": where there is a rank defficiency of the Jacobian

$$\begin{aligned} x&=(0,0,0,0) \quad &\text{with } f^T(x)f(x)&=2\times10^0 \\ x&=(1,2,2,2) \quad &\cdot \quad &=2.26\times10^0 \\ x&=(-1,-2,-2,-2) \quad &\cdot \quad &=4.1\times10^1 \end{aligned} \tag{4.5}$$

however, by automatically increasing the value of the parameter k at these points, the method was able to get the global solution.

5. CONCLUSIONS

In this paper another application of the tunneling concept to least square problems is presented. To arrive to the global zero residuals least squares solution of the problem, the Gauss-Newton method is used as the basis of the algorithm, and the tunneling mapping is employed to deal with singular or critical points for which the Gauss-Newton step Eq. (3.1) would not be defined. The numerical results clearly illustrate one of the basic properties of this method: if the pole strength k is sufficiently large the local solutions of the original problem are smoothed out and the Gauss-Newton displacements move towards the global solution.

We only outline here the basic ideas, a full description of the algorithm and the behaviour and sensitivity of the parameters can be found in Ref. [1].

6. AN IDEA IN PROGRESS

In section 3 we pointed out as one feature of the algorithm, the local stretching of the function, cancelling the critical points.

Another idea that presently is being explored, is that of a pole supported on a hypersphere.

In previous papers on global optimization, Ref. [2],[3],[4] the tunneling function has been used to deflate unwanted local minima of a function f(x) at x* using the expression

$$T(x) = \frac{f(x)-f(x^*)}{[(x-x^*)^T(x-x^*)]^k} \tag{6.1}$$

and during the tunneling phase a zero of T(x) is sought, to get a starting point of the next minimization phase.

If the solution of T(x)=0 is not found within a specified CPU time, the assumption is taken that probably the global minimum has been found at x* and the global optimization algorithm terminates.

Obviously this is only a necessary but not a sufficient condition for global optimality.

In order to increase the confidence, that really a solution of T(x)=0 does not exist, on the basis of a finite CPU time allocation, and idea that seems promising is to use a different mapping function instead of Eq. (6.1), defined by

$$\tau(x) = \frac{T(x)}{[R^2-(x-x^*)^T(x-x^*)]^k} \tag{6.2}$$

We note that in Eq. (6.2) a smoothing effect on T(x) accurs not by the action of a single pole at x* (as it was using Eq. (6.1)), but by a region of poles located at the boundary of the hypersphere of radius R.

This smoothing effect can easily be seen in Figs. 1 and 2, where the zero of the pulse like function is preserved by the mapping Eq. (6.2), and yet the function has been smoothed within the interior of the hypersphere increasing tremendously the zone of attraction of the zero.

Obviously, we are expressing here only the concept of an "idea in progress" and for conclusive numerical results, we shall have to wait for the next IIMAS workshop.

REFERENCES

[1] Gómez, S., Levy, A.V., Calderon, A., Cortés A., "The tunneling algorithm applied to zero residual least squares problems", Comunicaciones Técnicas del IIMAS-UNAM, Serie Naranja, No. 370, 1984.

[2] Levy, A.V., Montalvo, A., "The tunneling algorithm for the global minimization of functions", SIAM J. Sci. Stat. Comput. Vol. 6, No.1, January 1985.

[3] Levy, A.V., Gómez, S., "The tunneling algorithm for the global optimization of constrained functions", Comunicaciones Técnicas del IIMAS-UNAM, Serie Naranja, No. 231, 1980.

[4] Gómez, S., Levy, A.V., "The tunneling method for solving the constrained global optimization problem with several no-connected feasible regions", Lecture Notes in Maths., No. 909, Ed. J.P. Hennart, Springer-Verlag, 1981.

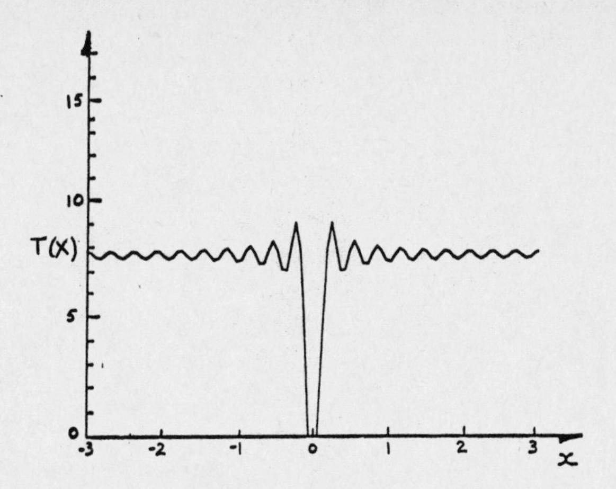

Figure 1a. Original pulse-like function, with zeroes near the origin and multiple singular points that cause a small region of attraction to the zeroes

$$T(x) = -10\left[\sum_{n=1}^{20}\left\{\frac{2}{n\pi}\sin(n\pi/36)\cos(xn)\right\}+\frac{2}{\pi}\sin(\pi/36)\right]+8$$

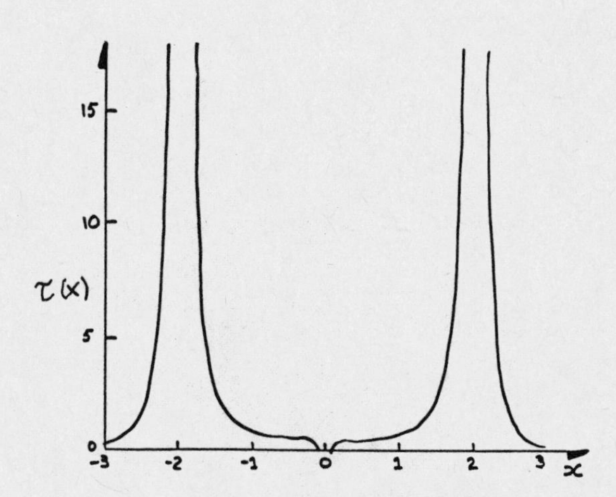

Figure 1b. Effect of the Mapping $\tau(x)$ on $T(x)$, Eq. (6.2) with R=2, k=2, $x^*=0$. The zeroes near the origin are preserved, while the singular points are smoothed, causing the region of attraction of the zeroes to increase.

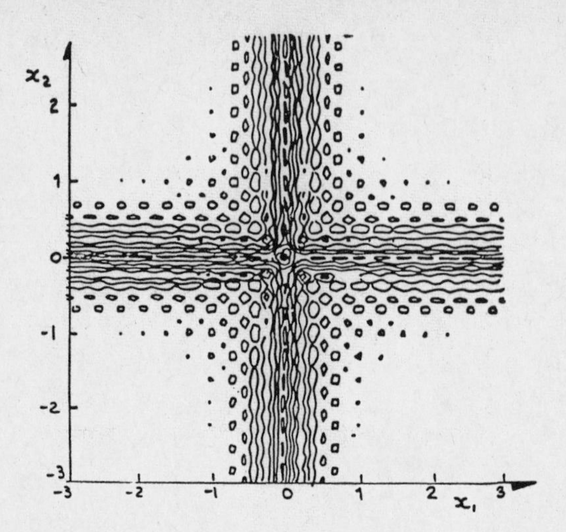

Figure 2a. Isolines of the original function T(x), with a zero at the origin and multiple singular points, that cause a small region of attraction to the zero

$$T(x) = \sum_{i=1}^{2} -10\left[\sum_{n=1}^{20} \left\{\frac{2}{n\pi}\sin(n\pi/36)\cos(nx_i)\right\} + \frac{2}{\pi}\sin(\pi/36)\right] + 16$$

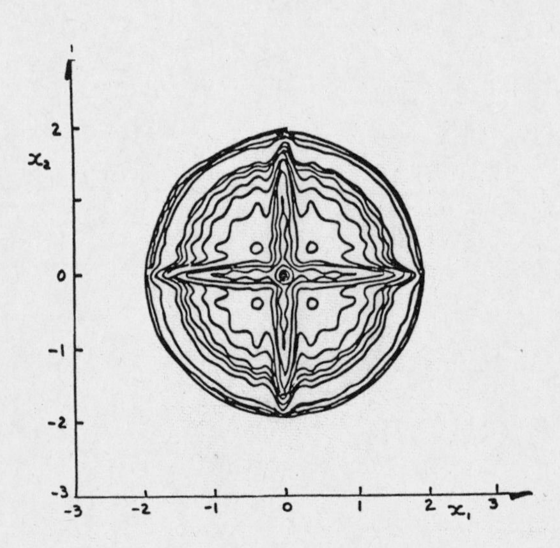

Figure 2b. Effect of the Mapping τ(x) on T(x), with R=2, k=1, x*=(0,0). The zero at the origin is preserved while the singular points are smoothed inside the circle, causing the region of attraction of the zero to increase.

EFFICIENT PRIMAL ALGORITHMS FOR STRICTLY CONVEX QUADRATIC PROGRAMS*

Donald Goldfarb
Department of Industrial Engineering
and Operations Research
Columbia University
New York, New York 10027

ABSTRACT

Two active set primal simplex algorithms for solving strictly convex quadratic programs are presented which, in their implementation, are closely related to the dual algorithm of Goldfarb and Idnani. Techniques are used for updating certain matrix factorizations that enable the algorithms to be both efficient and numerically stable in practice. One of the algorithms is based upon sufficient conditions for simultaneously dropping several constraints from the active set. It is shown how these conditions can be checked with little additional computational effort.

1. Introduction

In this paper, we consider the strictly convex quadratic programming (QP) problem

$$\underset{x}{\text{minimize}} \qquad f(x) = f_0 + c^T x + \tfrac{1}{2}x^T G x \qquad \text{(1a)}$$

$$\text{subject to} \quad s_j(x) \quad n_j^T x - b_j \geq 0, \quad j=1,\ldots,m \qquad \text{(1b)}$$

where x, c, and n_j, $j=1,2,\ldots,m$ are n-vectors, the constraint right hand sides b_j and slacks $s_j(x)$, $j=1,2,\ldots,m$ are scalars, G is an n x n symmetric positive definite matrix, and superscript T denotes transposition. In many QP problems there are equality constraints as well as inequality constraints. We shall ignore the former, however, for the sake of simplicity. Handling them directly requires only minor modifications to our algorithms.

The algorithms that we present here are "primal" (or "feasible point") "active-set" methods. In such methods each iterate $x^{(k)}$ always satisfies all of the constraints (1b)--i.e., $x^{(k)}$ is feasible--and the direction of movement from one iterate to the next is determined by the minimizer of f(x) over some subset (the active set) of the constraints in (1b), which are temporarily treated as equalities.

Many primal active-set methods have been proposed for QP; see [9] for references to these and other QP methods. Why then should one be interested in another primal active-set method? Our interest actually arose out of our work on a dual active-set method [9]. That method is very efficient and has been found to be very satisfactory when compared to other QP methods. (See the numerical results of Powell [10]

---

*This research was supported in part by the Army Research Office under Contract No. DAAG29-83-K-0106 and in part by the National Science Foundation under Grant No. DCR-83-41408.

in addition to those given in [9].) Moreover, we believed that the implementational aspects, which made that method efficient, could be transferred to a primal method. As this is indeed the case, several other justifications for developing the primal algorithms that we shall present here follow.

First, it is desirable to have a QP code which can solve problems by either primal or dual approaches. Clearly, for efficiency (in space and time) and ease of maintenance, both the primal and dual parts of such a code (or package of codes) should be based upon the same factorizations and be similar in approach so that they can share most of the same modules. If this is the case, then one can easily implement primal-dual QP algorithms as well. One such algorithm is described in [8] and many others are possible. We note that the computational results for the primal-dual algorithm reported in [8] would have been greatly improved had the primal part of that algorithm been similar to the dual part. Finally, sometimes one has to solve a sequence of problems in which there are either additions to or deletions from the set of constraints from one problem to the next or the right hand sides of the constraints are varied. Since these situations can require primal or dual steps to restore feasibility and optimality, it is useful to have both primal and dual capabilities.

In the next section, we outline the basic approach followed by our primal active set algorithms and we prove a lemma which underlies their implementation. Our first algorithm is presented in section 3. In that algorithm a single constraint is dropped when the constrained minimizer in the manifold (face) corresponding the current active set is reached and that point is not the optimal solution of the QP. An algorithm that allows several constraints to be dropped simultaneously is presented in section 4. That algorithm also allows constraints to be dropped at points that are not constrained minimizers. Techniques for updating certain matrix factorizations and vectors that enable the algorithms to be both efficient and numerically stable in practice are given in section 3. Simple examples which illustrate certain special features of our algorithms are also presented in sections 3 and 4.

2. Basic Approach and Relationships

Let $K=\{1,2,...,m\}$ denote the set of indices of the constraints (1b). By an "active set" (the term "working set" is used by some authors), we mean a subset of the constraints in (1b) which we temporarily require to be satisfied as equalities. We denote an active set, or to be more precise, the indices of the constraints in an active set, by A and we require the constraint normals $\{n_j;\ j\varepsilon A\}$ to be linearly independent. If we let $k \equiv |A|$ and denote the n x k matrix of these constraint normals by N then it is well-known that the vector x_A which minimizes the quadratic objective function (1a) subject to the equations

$$s_j(x) = n_j^T x - b_j = 0, \qquad j\varepsilon A \tag{2}$$

is uniquely determined by these equations and the condition that the gradient $g(x) \equiv \nabla f(x)$ of $f(x)$ at x_A be in the column space of N; i.e.,

$$g(x_A) = Nu_A \ . \tag{3}$$

u_A is the vector of k Lagrange multipliers (dual variables) corresponding to the active set A. x_A is the optimal solution to problem (1) if it is both primal and dual feasible; ie., x_A satisfies all constraints in (1b) and the u_A, uniquely determined by (3), is nonnegative.

Most primal QP algorithms proceed as follows.

Standard Primal Active Set Approach

(0) Assume that some active set A and corresponding constrained minimizer x_A, which is feasible, are given.

(1) If the dual variables u_A defined by (3) are nonnegative, stop; x_A is optimal. Otherwise choose a constraint corresponding to a negative dual variable and drop it from the active set.

(2) Compute the step z to the new x_A and move along the direction z to whichever is reached first, the new x_A (a full step) or the first point beyond which feasibility would be violated (a partial step). If a partial step is taken, add the limiting constraint to the active set and repeat this step. Otherwise go to (1).

There are many ways to implement the above approach. Some of the earliest QP methods that were proposed used tableaux (e.g. see Dantzig [2] and van de Panne and Whinston [11]). Subsequently, methods based upon projection operators (e.g., see Goldfarb [7] and Fletcher [4]) and methods based upon matrix factorizations (e.g. see Gill and Murray [5]) were developed.

Not all primal QP algorithms or even all primal active-set QP algorithms fit into the above mold. For example, Beale's method attempts to go from one constrained minimizer to the next along a sequence of mutually conjugate directions all of which lie in the null space of N^T corresponding to the current active set.

The above approach can also be generalized to allow the simultaneous dropping of more than one constraint from the active set A in step (1) and to allow additional constraints to be dropped from A just after a constraint has been added to A in step (2). One of the algorithms given in [7] is just of this type. A new and efficient implementation of it will be presented in section 4.

Before proceeding, we need to introduce some operators that are fundamental to active set algorithms for QP. For a given active set A, these are

$$N^* = (N^T G^{-1} N)^{-1} N^T G^{-1}, \tag{4}$$

the Moore-Penrose generalized inverse of N in the space of variables obtained under the transformation $y = G^{1/2}x$, and

$$H = G^{-1}(I - NN^*), \tag{5}$$

the reduced inverse Hessian for f(x) subject to the equality constraints (2).

As was stated earlier, we chose to implement our primal algorithms in a way that was as close as possible to the implementation of the dual algorithm of Goldfarb and Idnani. Consequently, instead of computing N^* and H explicitly, we store and update (when the active set changes) the matrices $J = L^{-T}Q$ and R, which are defined by the Cholesky factorization

$$G = LL^T$$

and the QR factorization

$$L^{-1}N = Q\begin{bmatrix} R \\ 0 \end{bmatrix} = [Q_1 \mid Q_2] \begin{bmatrix} R \\ 0 \end{bmatrix}.$$

L is (n x n) lower triangular, R is (k x k) upper triangular, and Q_1 and Q_2 consist of the first k and last n-k columns, respectively, of the orthogonal matrix Q. Partitioning J in the same manner as Q, i.e.,

$$J = [J_1 \mid J_2] = [L^{-T}Q_1 \mid L^{-T}Q_2],$$

we have that

$$N^* = R^{-1}J_1^T$$

and

$$H = J_2 J_2^T .$$

As in the dual algorithm of Goldfarb and Idnani [9], we shall require vectors of the form $z = Hn$ and $r = N^*n$. Given J and R it follows directly from the above that these can be computed as

$$z = J_2 d_2 \quad \text{and} \quad r = R^{-1}d_1 \tag{6}$$

where

$$d = \begin{bmatrix} d_1 \\ \hline d_2 \end{bmatrix} = \begin{bmatrix} J_1^T \\ \hline J_2^T \end{bmatrix} n = J^T n. \tag{7}$$

Techniques for updating J, R, and d when A changes are described in section 3.

Let us now turn to another basic idea behind our implementation and that of [9] as well. In our algorithms we move along a broken-line path from one constrained minimum x_{A_1} to the next x_{A_2}. The breaks (i.e., changes in direction) occur where the active set changes. The key to our approach is that until x_{A_2} is reached, we continue to express the gradient of the objective function as a linear combination of <u>all</u> of the normals to the constraints that were active at x_{A_1} (i.e., those in A_1). Geometrically, this corresponds to parametrically adjusting the right hand sides of the inequalities with indices in $A_1 \setminus A_2$ so that these constraints are satisfied as equalities all along the path from x_{A_1} to x_{A_2}. We need to know how this expression, or equivalently, how the vector of dual variables u, varies along this path, and we need to allow constraints to be added to and dropped from the active set along this path.

This is the essence of the following basic lemma.

<u>Basic Lemma</u>

Given the sets A and $A^+ = A \cup \{p_1,\ldots,p_j\}$, let x be a point such that

$$s_i(x) = 0, \qquad \text{for all } i \varepsilon A \tag{8}$$

and

$$g(x) = N^+u^+ = \sum_{i \varepsilon A} u_i n_i + \hat{n} \tag{9}$$

where

$$\hat{n} = \sum_{i=1}^{j} u_{p_i} n_{p_i} . \tag{10}$$

Then for all points of the form

$$\overline{x}(t) = x + tz \tag{11}$$

where

$$z = -H\hat{n} , \tag{12}$$

we have that

$$s_i(\overline{x}(t)) = 0, \qquad \text{for all } i \varepsilon A \tag{13}$$

$$g(\overline{x}(t)) = N^+\overline{u}^+(\overline{x}(t)) \tag{14}$$

where

$$\overline{u}^+(\overline{x}(t)) = u^+ + t \begin{bmatrix} r \\ -u_{p_1} \\ \vdots \\ -u_{p_j} \end{bmatrix} \tag{15}$$

and

$$r = N^*\hat{n}. \tag{16}$$

Proof: (13) follows from (8), (11) and (12) and the fact that HN = 0. To prove (14) and (15) we note that

$$Gz = -GH\hat{n} = -(I-NN^*)\hat{n} = Nr - \hat{n} = N^+ \begin{bmatrix} r \\ -u_{p_1} \\ \vdots \\ -u_{p_j} \end{bmatrix}.$$

Hence, $$g(\bar{x}(t)) = g(x) + tGz = N^+(u + t\begin{bmatrix} r \\ -u_{p_1} \\ \vdots \\ -u_{p_j} \end{bmatrix}) = N^+\bar{u}^+(\bar{x}(t)).$$

To demonstrate how this lemma plays a fundamental role in our algorithms, suppose that we are at a constrained minimizer x_{A_1} correspond- to the active set $A_1 \equiv A^+ \equiv A \cup \{p_1,\ldots,p_j\}$. Suppose now that we drop the constraint indices $p_1,\ldots,p_j$ from the active set. Clearly, the point x_{A_1} satisfies the conditions of the lemma; hence it follows from the lemma that the constrained minimum x_A is given by

$$x_A = x_{A_1} + z$$

where

$$z = -H\hat{n} = -Hg(x_{A_1}). \qquad (17)$$

Observe that setting t=1 in the lemma causes the last k components of $\bar{u}^+(\bar{x}(1))$ to vanish. Consequently, $g(x_A)$ has the form $g(x_A) = Nu_A$ and $s_i(x_A)=0$ for all $i\varepsilon A$, where u_A is the vector consisting of the first k components of $\bar{u}^+(\bar{x}(1))$. This shows that x_A is indeed the constrained minimum corresponding to A.

Also we note that the equivalence of the two expressions for z in (17) follows from (9) and the fact that HN=0.

Now suppose that taking a full step z to x_A is not possible because it would violate some constraints in (1b). Let constraint q be the first constraint that is violated by moving in the direction z, and consider the point

$$\bar{x}(\hat{t}) = x_{A_1} + \hat{t}z$$

where $s_q(\bar{x}(\hat{t}))=0$ and, of course, $\hat{t} < 1$.

At this point, let us add q to the active set A, i.e., we set $A \leftarrow A \cup \{q\}$ and $A^+ \leftarrow A^+ \cup \{q\}$. If we let q be the first index in these sets so that $N \leftarrow [n_q \,\vdots\, N]$ and we set

$$u^+ = \bar{u}^+(\bar{x}(\hat{t})) \leftarrow \begin{bmatrix} 0 \\ \bar{u}^+(\bar{x}(\hat{t})) \end{bmatrix},$$

then clearly the point $x = \bar{x}(\hat{t})$ satisfies the conditions of the lemma with respect to the new sets A and A^+ as well as the old. Since the dual variables corresponding to the constraints in $A^+ \setminus A$, i.e., $u_{p_1}, \ldots, u_{p_j}$, have been reduced to $(1-\hat{t})$ times their value at the start of the step, we need only to replace $\hat{n}$ by $(1-\hat{t})\hat{n}$, compute $z = -H\hat{n}$ and $r = N^*\hat{n}$, using H and N^* corresponding to the new active set A, and continue moving in the new direction z towards the new constrained minimizer x_A.

3. A Primal Active-Set QP Algorithm

We are now ready to present a primal active-set QP algorithm that follows the approach outlined in the previous section. Its implementation is as described there except that the vector d, which is required in (6) to compute the directions of movement in the primal and and dual spaces is updated after a change is made to the active set rather than computed afresh from (7). These efficiencies are described after the presentation of the algorithm. Also, we note that it is not necessary to update the dual variables corresponding to indices in $A^+ \setminus A$.

Algorithm 1:

0) Find a feasible constrained minimizer x

Set A to the active set, and compute the primal slacks s(x), the matrices J and R (given the Cholesky factor L) and the dual variables $u = R^{-1} J_1^T g(x)$.

1) Check for optimality -- choose constraint to drop

If $V \equiv \{j \in A \mid u_j < 0\} = \emptyset$, STOP; x is optimal. Otherwise, choose $p \in V$ to drop from the active set. Set $A \leftarrow A \setminus \{p\}$, $d \leftarrow u_p v$, where v is the column of $\begin{bmatrix} R \\ 0 \end{bmatrix}$ corresponding to n_p, and update J, R, and d.

2) Compute step direction and length

a) Compute step directions in primal and dual spaces.

Compute $z = -J_2 d_2$, if $k < n$, (else set z=0),

and $r = R^{-1} d_1$, if $k > 0$.

b) Compute step length.

(i) Maximum step length without violating primal feasibility:

If $v_j \equiv n_j^T z \geq 0$, for all $j \in K \setminus A$, set $t_1 \leftarrow \infty$;

otherwise set $t_1 \leftarrow \min\{-s_j/v_j \mid v_j < 0, j \in K \setminus A\}$.

(ii) Actual step length:

Set $t \leftarrow \min\{t_1, 1\}$.

c) Take steps in primal and dual spaces.

Set $x \leftarrow x + tz$,

$u \leftarrow u + tr$,

and $s_j \leftarrow s_j + tv_j$, for all $j \in K \setminus A$.

(Note: z may equal 0 even if $r \neq 0$.)

If $t < 1$ (partial step), add q to the active set as the last constraint; i.e., set $A \leftarrow A \cup \{q\}$, update J, R, and d, set $u \leftarrow \binom{u}{0}$, $d \leftarrow (1-t)d$ and go to (2).

If $t = 1$ (full step), go to (1).

To complete the description of this algorithm, we need to specify how to update the matrices J and R, and the vector d. Since we use the same techniques for updating J and R as those described in [9], we shall be quite brief. In what follows $\hat{Q}$ will represent the product of a sequence of appropriately chosen Givens plane rotation matrices.

Consider adding q to A to give the new active set $\bar{A} = A \cup \{q\}$. If $\hat{Q}$ is chosen so that

$$\hat{Q}h_2 = \gamma e_1,$$

where $\gamma = \pm \|h\|_2$ and h_2 is the vector of the last n-k components of

$$h = \begin{pmatrix} h_1 \\ h_2 \end{pmatrix} = J^T n_q ,$$

then since

$$\begin{bmatrix} I & 0 \\ 0 & \hat{Q} \end{bmatrix} J^T (N \mid n_q) = \begin{pmatrix} I & 0 \\ 0 & \hat{Q} \end{pmatrix} \left[\begin{array}{c|c} R & h_1 \\ \hline 0 & h_2 \end{array}\right] = \left[\begin{array}{c|c} R & h_1 \\ \hline 0 & \gamma e_1 \end{array}\right]$$

we have that

$$\bar{J} = [\, \underbrace{\bar{J}_1}_{k+1} \mid \underbrace{\bar{J}_2}_{n-k-1} \,] = [\, \underbrace{J_1}_{k} \mid \underbrace{J_2 \hat{Q}^T}_{n-k} \,] \tag{18}$$

and

$$\bar{R} = \left[\begin{array}{c|c} R & h_1 \\ \hline 0 & \gamma \end{array}\right].$$

Moreover (18) and the definition of d imply that

$$(\, \underbrace{\bar{d}_1^{\,T}}_{k+1} \mid \underbrace{\bar{d}_2^{\,T}}_{n-k-1} \,) = (\, \underbrace{d_1^T}_{k} \mid \underbrace{d_2^T \hat{Q}^T}_{n-k} \,)$$

Consider now dropping p from A^+ to give A, and assume that n_p is the $\hat{k}$-th column of N^+. Consequently,

$$(J_1^+)^T N = \begin{bmatrix} R_1 & S \\ 0 & T \end{bmatrix},$$

where the matrix on the right hand side of the above equation is R^+ with its $\hat{k}$-th column deleted. If $\hat{Q}$ is chosen so that

$$\hat{Q}T = \begin{bmatrix} R_2 \\ 0 \end{bmatrix}$$

where R_2 is a $(k+1-\hat{k}) \times (k+1-\hat{k})$ upper triangular matrix and $\hat{Q}$ is $(k+2-\hat{k}) \times (k+2-\hat{k})$, then

$$R = \begin{bmatrix} R_1 & S \\ 0 & R_2 \end{bmatrix}, \quad J = J^+ \begin{bmatrix} \tilde{Q}^T & 0 \\ 0 & I \end{bmatrix} = [\underbrace{J_1^+ \tilde{Q}^T}_{k+1} \mid \underbrace{J_2^+}_{n-k-1}],$$

and

$$(\underbrace{d_1^T}_{k} \mid \underbrace{d_2^T}_{n-k}) = (\underbrace{(d_1^+)^T \tilde{Q}^T}_{k+1} \mid \underbrace{(d_2^+)^T}_{n-k-1}),$$

where

$$\tilde{Q} = \begin{bmatrix} I & 0 \\ 0 & \hat{Q} \end{bmatrix}.$$

Moreover, since $(J^+)^T n_p = \begin{bmatrix} R^+ \\ 0 \end{bmatrix} e_{\hat{k}}$, it follows that $d_1^+ = u_p R^+ e_{\hat{k}}$ and $d_2^+ = 0$ and that d_1 and d_2 in step (1a) of algorithm 1 can be computed as

$$\begin{pmatrix} d_1 \\ \delta \end{pmatrix} = \tilde{Q}(u_p R^+ e_{\hat{k}}) \text{ and } d_2 = \delta e_1.$$

It also follows that z in step (2a) is given simply as

$$z = -\delta J_2 e_1 .$$

Also, observe that the above computations are equivalent to cyclically permuting the last $k+2-\hat{k}$ columns of R^+ so that column $\hat{k}$ becomes column $\hat{k}+1$, followed by a reduction of the resulting upper Hessenberg matrix to the upper triangular matrix

$$\begin{bmatrix} R & d_1 \\ 0 & \delta \end{bmatrix}$$

using Givens rotations.

In order to estimate the amount of work that is required by algorithm 1, we shall assume that a typical iteration involves one deletion from and one addition to the active set. We shall also assume that all matrices are dense and that algorithms described in Daniel et al. [1] are used for computing and applying Givens rotations. The approximate operational counts for the major computational steps of algorithm 1 are given in Table 1. An "operation" is one multiplication and one addition and we include only terms that are quadratic in n, m, and k. The number of operations given for updating J, d, and R when a constraint is dropped was obtained by averaging over the k values that $\hat{k}$ (the position in A of the dropped constraint) could take.

Table 1: Computational Effort of a Typical Iteration of Algorithm 1:

Computational Step	Operations
Compute $v_j = n_j^T z$	$n(m-k)$
Add:	
Update J, R, and d (includes $h=J^T n_q$)	$3n(n-k) + n^2$
Compute z and r	$n(n-k) + k^2/2$
Drop:	
Update J, R and d	$1/2\ k\ (3n+k)$
Compute z and r	$k^2/2$
Total Work: W	$nm + 5n^2 - 7/2\ nk + 3/2\ k^2$

Averaging the total work given in Table 1 over k gives an "average" operational count of

$$\bar{W} = nm + 3.75n^2$$

for Algorithm 1 per iteration. The work required by an iteration of algorithm 1 is essentially the same as that required by the dual algorithm of Goldfarb and Idnani [9], assuming that in both algorithms an index is added to A and another is dropped from A. The main difference between the operational counts given in Table 1 and those given for the Goldfarb-Idnani dual algorithm by Powell [10] comes from the fact that the formulas that we use for applying a Givens rotation to a vector (see, Daniel, et al. [1]) requires three multiplications and additions, while Powell uses a more obvious set of formulas which requires four multiplications and two additions.

We note that Powell [10] reports that the two general QP subroutines, QPSOL [6] and VE02A [3] require

$$W(QPSOL) = nm + 13.5n^2 - 22nk + 12k^2$$

and

$$W(VE02A) = nm + 6n^2 + 4nk$$

for a "typical" iteration. Averaging over k yields

$$\bar{W}(QPSOL) = nm + 6.5n^2$$

and

$$\bar{W}(VE02A) = nm + 8n^2 .$$

It should be stressed that except for their implementation, these subroutines follow the same basic approach as does algorithm 1. Thus, our claim regarding the efficiency of algorithm 1 appears to be justified. We note, however, that both VE02A and QPSOL are capable of solving positive semi-definite and indefinite QP's. In the latter case, only a local solution is obtained.

An interesting aspect that algorithm 1 shares with the Goldfarb-Idnani dual algorithm is that "purely" dual steps are possible, i.e., u changes but x does not. To illustrate this, consider the problem

$$\text{minimize } \frac{1}{2}x_1^2 + \frac{1}{2}x_2^2$$
$$\text{subject to } -x_2 \geq -2$$
$$x_2 \geq 1 .$$

Starting from the constrained minimizer $x=(0,2)^T$ corresponding to the active set $A=\{1\}$. Since $u_1 = -2$ we set $A \leftarrow A \setminus \{1\}=\emptyset$ and take a partial step $(t=1/2)$ to the point $x=(0,1)^T$, and set $A \leftarrow A \cup \{2\}=\{2\}$. Since this point is the constrained minimizer corresponding to $A=\{2\}$, $z=0$; however, $r=1$ and $t=1$ and a purely dual step is taken. It is easily seen that such steps will occur whenever a partial step is taken to a constrained minimizer, since in that case the dual variables u_i^+, for $i \varepsilon A^+ \setminus A$, have not yet been reduced to zero. This is then achieved by a single, full, purely dual step.

4. A Primal Active-Set Algorithm that Allows Multiple Drops

In [7], we proposed a primal active-set QP algorithm that allowed several constraints to be simultaneously dropped from the active set at a constrained minimizer and that allowed additional constraints to be dropped after a partial step. The basis for that algorithm was a theorem giving sufficient conditions for simultaneously dropping several constraints, that depended upon recurrence formulas involving the optimal Lagrange multipliers (see (3)) for nested sets of active sets.

Let A_k denote the active set $\{1,2,...,k\}$, and let

$$u^k = N_k^* g(x^k) = N_k^T G^{-1} N_k)^{-1} N_k^T G^{-1} g(x^k),$$

where $N_k=[n_1,n_2,...,n_k]$ and x^k is <u>any</u> point that satisfies the constraints indexed by A_k as equalities, i.e.,

$$s_i(x^k) = n_i^T x^k - b_i = 0, \qquad \text{for all } i \varepsilon A_k . \tag{19}$$

We note that the definition of the optimal Lagrange multiplier vector u^k is independent of the choice of x^k so long as it satisfies (19). The theorem giving sufficient conditions for multiple drops that is proved in [7] can be stated as:

<u>Theorem</u> (<u>Sufficient Conditions for Multiple Drops</u>)

Let the point x^k satisfy all of the constraints (1b) and only those indexed by A_k as equalities (i.e., (19)). If

$$u_i^{j-1} \leq u_i^j , \qquad \begin{cases} q < j \leq k \\ q \leq i < j \end{cases} \tag{20}$$

and

$$u_j^j < 0 \ , \qquad q \le j \le k \ , \qquad (21)$$

then the direction $z = -H_{q-1}g(x^k)$ is feasible

where $H_{q-1} = G^{-1}(I-N_{q-1}N^*_{q-1})$.

Simply put, this theorem states that it is all right to drop a constraint from the active set if its optimal Lagrange multiplier is negative and this multiplier did not increase when the constraints that were dropped previously at the current point were dropped.

Our implementation of these sufficient conditions for simultaneously dropping several constraints in Algorithm 2 below is based upon the observation that

$$u_i^j = u_i^k + r_i^j \ , \qquad 1 \le i \le j < k \ ,$$

where

$$r^j = N_j^* \hat{n}$$

and

$$\hat{n} = \sum_{p=j+1}^{k} u_p^p n_p \ .$$

Consequently, (20) and (21) are equivalent, respectively, to

$$r_i^{j-1} \le r_i^j \ , \begin{cases} q < j \le k \\ q \le i < j \end{cases}$$

and

$$u_j^k + r_j^j < 0 \ , \qquad q \le j < k \ ,$$

where $r_i^k = 0$, $i=1,\ldots,k$.

Since r is the step direction in the dual space and is required in any case by our algorithm, it follows that an inconsequential amount of work is needed to check these sufficient conditions to <u>determine</u> if an additional constraint should be dropped. Dropping the constraint is, of course, a nontrivial computation.

<u>Algorithm 2</u>:

0) <u>Find a feasible constrained minimizer x</u>

Set A to the active set, and compute the primal slacks s(x), the matrices J and R and the dual variables $u = R^{-1}J_1^T g(x)$.

1) <u>Check for optimality</u>

If $V = \{j \varepsilon A \mid u_j < 0\} = \emptyset$, STOP; x is optimal.

Otherwise set $C \leftarrow A$, $D \leftarrow \emptyset$, $d \leftarrow 0$ and $r \leftarrow 0$.

(Note: D is the set of indices of constraints dropped since a constrained minimizer was last computed.)

2) <u>Choose constraints to drop</u>

Repeat until $V = \emptyset$.

Choose $p \varepsilon V \backslash D$ to drop from the active set.

Set $A \leftarrow A \backslash \{p\}$, $D \leftarrow D \cup \{p\}$, $C \leftarrow C \backslash \{p\}$, $d \leftarrow d + u_p v$ where v is the column of $[\begin{smallmatrix} R \\ 0 \end{smallmatrix}]$ corresponding to n_p, and update J, R, and d.

Compute the step direction in dual space $\bar{r} = R^{-1} d_1$, if $k > 0$, and set $C \leftarrow \{j \varepsilon C \mid \bar{r}_j \leq r_j\}$, $r \leftarrow \bar{r}$ and $V \leftarrow \{j \varepsilon C \mid u_j + r_j < 0\}$.

3) <u>Compute step direction and length</u>

a) Compute step direction in primal space.

Compute $z = -J_2 d_2$, if $k < n$, (else set $z=0$).

b) Compute step length.

(i) Maximum step without violating primal feasibility:

If $v_j \equiv n^T z \geq 0$ for all $j \varepsilon K \backslash A$, set $t_1 \leftarrow \infty$; otherwise set $t_1 \leftarrow \min \{-s_j/v_j \mid v_j < 0, j \varepsilon K \backslash A\} = -s_q/v_q$

(ii) Actual step length:

Set $t \leftarrow \min \{t_1, 1\}$

(iii) Take step in primal and dual spaces.

Set $x \leftarrow x + tz$

$u \leftarrow u + tr$

and $s_j \leftarrow s_j + tv$, for all $j \varepsilon K \backslash A$.

(Note: z may equal 0 even if $r \neq 0$.)

If $t < 1$ (partial step) add constraint q to active set as the last constraint; i.e., set $A \leftarrow A \cup \{q\}$, update J, R and d and set $u \leftarrow (\begin{smallmatrix} u \\ 0 \end{smallmatrix})$, $d \leftarrow (1-t)d$, $r \leftarrow R^{-1} d_1$, $C \leftarrow A$, and $V \leftarrow \{j \varepsilon A \mid u_j + r_j < 0\}$ and go to (2). If $t=1$ (full step) go to (1).

Notice that only those constraints in $V \backslash D$ are allowed to be dropped from the active set. This guarantees that the algorithm terminates in a finite number of iterations because a constraint that has been dropped from the active set, and subsequently added back to it, cannot be dropped again until a constrained minimizer is reached.

The example below shows that the above drop-add scenario can occur. It also illustrates the application of the sufficient condition theorem and shows that pure dual steps are possible in algorithm 2.

<u>Problem</u>: Minimize $1/2 \, (x_1^2 + x_2^2)$

Subject to
$$\begin{aligned} -x_1 \quad & \geq -1 \\ -x_1 - 2x_2 & \geq -7 \\ x_1 + x_2 & \geq 3 \end{aligned}$$

<u>Solution by Algorithm 2</u>:

Starting at the feasible vertex $x=(1,3)^T$ with $A=\{1,2\}$ compute $s^T=(0,0,1)$, $J=I$, $R=N=\begin{bmatrix}-1 & -1\\ 0 & -2\end{bmatrix}$ and $u=(\frac{1}{2},-\frac{3}{2})^T$.

<u>Iteration 1</u>:

Step (1): $V=\{2\}$. Set $C=\{1,2\}$, $D=\emptyset$, $d=0$ and $r=0$.

Step (2): Drop a constraint: Choose $p=2$. Set $C=A=\{1\}$, $D=\{2\}$,

$$d^T= -\frac{3}{2}(-1,-2)=(\frac{3}{2},3), \quad J=I, \text{ and } R=[-1].$$

Compute $\bar{r}= -\frac{3}{2}$ and set $C=\{1\}$, $r= -\frac{3}{2}$, $V=\{1\}$

(since $u_1+r_1 = -1 < 0$).

Drop a second constraint:

Choose $p=1$. Set $C=A=\emptyset$, $D=\{1,2\}$, $d^T=(1,3)$, and $J=I$.

(Note: $k=0$.) Set $C=V=\emptyset$.

Step (3): Compute

$$z^T = (-1,-3), \quad t=t_1= \frac{1}{4} \quad (q=3)$$

$$x^T = (1,3) + \frac{1}{4}(-1,-3) = (\frac{3}{4},\frac{9}{4})$$

$$s^T = (0,0,1) + \frac{1}{4}(1,7,-4) = (\frac{1}{4},\frac{7}{4},0).$$

Set $A=\{3\}$, $J=\frac{1}{\sqrt{2}}\begin{bmatrix}1 & -1\\ 1 & 1\end{bmatrix}$, $R=[\sqrt{2}\ ]$

$$d^T = (1 \frac{1}{\sqrt{2}} (4,2), \quad u=[0].$$

Compute $d^T = (1 - \frac{1}{4}) \frac{1}{\sqrt{2}} (4,2) = \frac{1}{\sqrt{2}} (3,\frac{3}{2})$, $r=\frac{3}{2}$.

$C=\{3\}$, and $V=\emptyset$ (since $u_1 + r_1 = \frac{3}{2} > 0$.)

<u>Iteration 2</u>:

Step (2): Since $V=\emptyset$, we proceed to step 3.

Step (3): Compute

$$z^T = \frac{-1}{\sqrt{2}} (\frac{3}{2}) \frac{1}{\sqrt{2}} (-1,1) = (\frac{3}{4},-\frac{3}{4})$$

$$t = t_1 = (-\frac{1}{4} /- \frac{3}{4}) = \frac{1}{3} \quad (q=1)$$

$$x^T = (\frac{3}{4},\frac{9}{4}) + \frac{1}{3} (\frac{3}{4},-\frac{3}{4}) = (1,2)$$

$$u = 0 + \frac{1}{3} (\frac{3}{2}) = \frac{1}{2}$$

$$s^T = (\frac{1}{4},\frac{7}{4},0) + \frac{1}{3}(-\frac{3}{4},\frac{3}{4},0) =(0,2,0)$$

(Note: Constraint $q=1$ which was previously dropped is now added back to the active set.)

Set $A = \{3,1\}$, $J = \frac{1}{\sqrt{2}} \begin{bmatrix} 1 & -1 \\ 1 & 1 \end{bmatrix}$, $R = \frac{1}{\sqrt{2}} \begin{bmatrix} 2 & -1 \\ 0 & 1 \end{bmatrix}$,

$u^T = (\frac{1}{2},0)$, $d^T = (1-\frac{1}{3}) \frac{1}{\sqrt{2}}(3,\frac{3}{2}) = \frac{1}{\sqrt{2}} (2,1)$,

$r^T = (\frac{3}{2},1)$. $C=\{3,1\}$ and $V=\emptyset$ (since $u^T + r^T = (2,1) > 0$).

Iteration 3:

Step (2): Since $V=\emptyset$, we proceed to step 3.

Step (3): Since $k=n$, this is a pure dual step; $z=0$, $t_1=\infty$, $t=1$, and $x^T = (1,2)$ and $s^T = (0,2,0)$ do not change.

Compute

$$u^T = (\tfrac{1}{2},0) + (\tfrac{3}{2},1) = (2,1).$$

Iteration 4:

Step (1): Since $V=\emptyset$; STOP, $x=(1,2)$ is the optimal solution.

REFERENCES

1. J. W. Daniel, W. B. Gragg, L. Kaufman and G. W. Stewart."Reorthogonalization and stable algorithms for updating the Gram-Schmidt QR factorizations." Mathematics of Computation 30 (1976) 772-795.
2. G. B. Dantzig. Linear programming and extensions (Princeton University Press, Princeton, N.J. (1963) Chapter 24, Section 4.
3. R. Fletcher. "A FORTRAN subroutine for quadratic programming." UKAEA Research Group Report. AERE R6370 (1970).
4. R. Fletcher. "A general quadratic programming algorithm." Journal of the Institute of Mathematics and Its Applications (1971) 76-91.
5. P. E. Gill and W. Murray. "Numerically stable methods for quadratic programming." Mathematical programming 14 (1978) 349-372.
6. P. E. Gill, W. Murray, M. A. Saunders and M. H. Wright. "User's guide for SOL/QPSOL: a Fortran package for quadratic programming." Report SOL 83-7 (Stanford University, 1983).
7. D. Goldfarb. "Extension of Newton's method and simplex methods for solving quadratic programs," in: F. A. Lootsma, ed., Numerical methods for nonlinear optimization (Academic Press, London, 1972) 239-254.
8. D. Goldfarb and A. Idnani. "Dual and primal-dual methods for solving strictly convex quadratic programs," in: J. P. Hennart, ed., Numerical Analysis, Proceedings Cocoyoc, Mexico 1981. Lecture Notes in Mathematics 909 (Spring-Verlag, Berlin, 1982) 226-239.
9. D. Goldfarb and A. Idnani, "A numerically stable dual method for solving strictly convex quadratic programs." Math. Programming 27 (1983) 1-33.
10. M. J. D. Powell. "On the quadratic programming algorithm of Goldfarb and Idnani." Report DAMTP 1983/NA19 (University of Cambridge, 1983).
11. C. Van de Panne and A. Whinston. "The simplex and the dual method for quadratic programming." Operations Research Quarterly 15 (1964) 355-389.

LOCATION OF MULTIPLE EQUILIBRIUM CONFIGURATIONS NEAR LIMIT POINTS BY A DOUBLE DOGLEG STRATEGY AND TUNNELLING

L.T. Watson*
Department of Computer Science
Virginia Polytechnic Inst. and State Univ.
Blacksburg, VA 24061 USA

M.P. Kamat
Dept. of Engng. Science and Mech.
Virginia Polytechnic Inst. and State Univ.
Blacksburg, VA 24061 USA

H.Y. Kwok*
Department of Computer Science
Virginia Polytechnic Inst. and State Univ.
Blacksburg, VA 24061 USA

Abstract

A hybrid method for locating multiple equilibrium configurations has been proposed recently. The hybrid method combined the efficiency of a quasi-Newton method capable of locating stable and unstable equilibrium solutions with a robust homotopy method capable of tracking equilibrium paths with turning points and exploiting sparsity of the Jacobian matrix at the same time. A quasi-Newton method in conjunction with a deflation technique is proposed here as an alternative to the hybrid method. The proposed method not only exploits sparsity and symmetry, but also represents an improvement in efficiency. Limit points and nearby equilibrium solutions, either stable or unstable, can be accurately located with the use of a modified pseudoinverse based on the singular value decomposition. This pseudoinverse modification destroys the Jacobian matrix sparsity, but is invoked only rarely (at limit and bifurcation points).

Introduction

In predicting response of structures susceptible to limit and bifurcation point instabilities, previous techniques as in [1] to [4], suffered serious difficulties in the vicinity of limit points. The present algorithm is proposed to overcome these difficulties, and is successful in locating equilibrium solutions in the vicinity of a limit point to great accuracy. The algorithm extends a quasi-Newton method with a deflation technique to solve the system of nonlinear equilibrium equations directly; multiple equilibrium solutions (stable or unstable), if they exist, can be located efficiently.

*Supported in part by NSF Grant MCS#8207217

From a recent evaluation by Kamat, Watson, and Venkayya [5], the globally convergent quasi-Newton method of Dennis and Schnabel [6], although quite efficient, has its limitations in the vicinity of limit and bifurcation points and along unloading portions of the equilibrium curve, especially when used in the context of energy minimization. Since the Jacobian matrix of the system of nonlinear equilibrium equations is nearly singular in the vicinity of a critical (limit or bifurcation) point, quasi-Newton iterations encounter serious numerical difficulties. Gay [7] suggests using a modified pseudoinverse in place of the inverse of the Jacobian matrix in the Newton iteration to maintain numerical stability.

After locating the first equilibrium solution at a certain fixed load level, deflation is used to locate multiple stable and unstable equilibrium solutions which may exist. The equilibrium solutions already found are used to deflate a nonlinear least squares function, which is used with a model trust region quasi-Newton algorithm to find another equilibrium solution (stable or unstable). If another equilibrium solution exists, it can be located in a finite number of iterations. In minimizing the deflated function, the quasi-Newton method can proceed using the Jacobian matrix of the previous (undeflated) nonlinear least squares function while continuing to exploit sparsity and symmetry.

By means of the matrix factorization LDL^t, the algorithm checks the condition number of the Jacobian matrix of the system of nonlinear equations at every iteration. The Jacobian matrix becoming ill-conditioned is generally an indication of entering the vicinity of a critical point, in which case Gay's modification [7] is used to perturb the Jacobian matrix into a better conditioned one to maintain numerical stability.

From any starting point, the globally convergent quasi-Newton method may converge to a local minimum or fail to make reasonable progress. The proposed algorithm detects this situation from the gradient norm or from the fact that the algorithm is not making reasonable progress with good directions. Local minima are used to deflate the nonlinear least-squares function by a procedure known as tunnelling [8-12], and the ultimate result of tunnelling yields an equilibrium solution of the load level. So, the proposed method usually is "globally convergent".

Model Trust Region Quasi-Newton Method

Let R^n denote n-dimensional Euclidean space, and

$$F : R^n \longrightarrow R^n$$

be twice continuously differentiable. The problem is to find $X_* \in R^n$ such that

$$F(X_*) = 0 \quad . \tag{1}$$

A double dogleg strategy is applied to minimize the nonlinear least squares function:

$$f(X) = 1/2F^t(X)F(X) \quad . \tag{2}$$

Of course, a local minimum of the nonlinear least squares function may not be a solution of the simultaneous nonlinear equations. Special techniques, such as tunnelling, have to be used to force the solution of the nonlinear least squares function away from local minima.

A quadratic model

$$m_c(X) = f(X_c) + \nabla f(X_c)^t(X-X_c) + 1/2(X-X_c)^t H_c(X-X_c)$$

is built around the current estimate X_c. A step $S=(X_+-X_c)$ is taken to minimize $m_c(X)$ within a region of radius δ_c, where the quadratic model can be "trusted". H_c is the Hessian at X_c and is approximated by J^tJ, where J is the Jacobian matrix of F evaluated at X_c.

If the Newton step

$$S_c^N = -H_c^{-1}\nabla f(X_c) \tag{4}$$

is within the trust region, then $X_+=X_c+S_c^N$ is taken as the next point since it is the global minimum of the model. Otherwise, the minimizer of the quadratic model occurs for

$$S = S(\mu) \equiv -(H_c+\mu I)^{-1}\nabla f(X_c) \ , \ \mu \geqslant 0 \ ,$$

$$\text{such that } ||S(\mu)||_2 = \delta_c \quad .$$

The $S(\mu)$ curve, as shown in Figure 1, runs smoothly from the Newton step when $\mu=0$ to the steepest descent direction

$$S(\mu) \simeq -1/\mu \ \nabla f(X_c)$$

for large μ.

The double dogleg strategy is to approximate the $S(\mu)$ curve by the double dogleg arc which connects the "Cauchy point" to a point $X_{\hat{N}}$ in the Newton direction for m_c, and choose X_+ to be the point on this arc such that $||X_+-X_c||=\delta_c$. The strategy looks in the steepest descent direction when δ_c is small and more and more towards the Newton direction as δ_c increases.

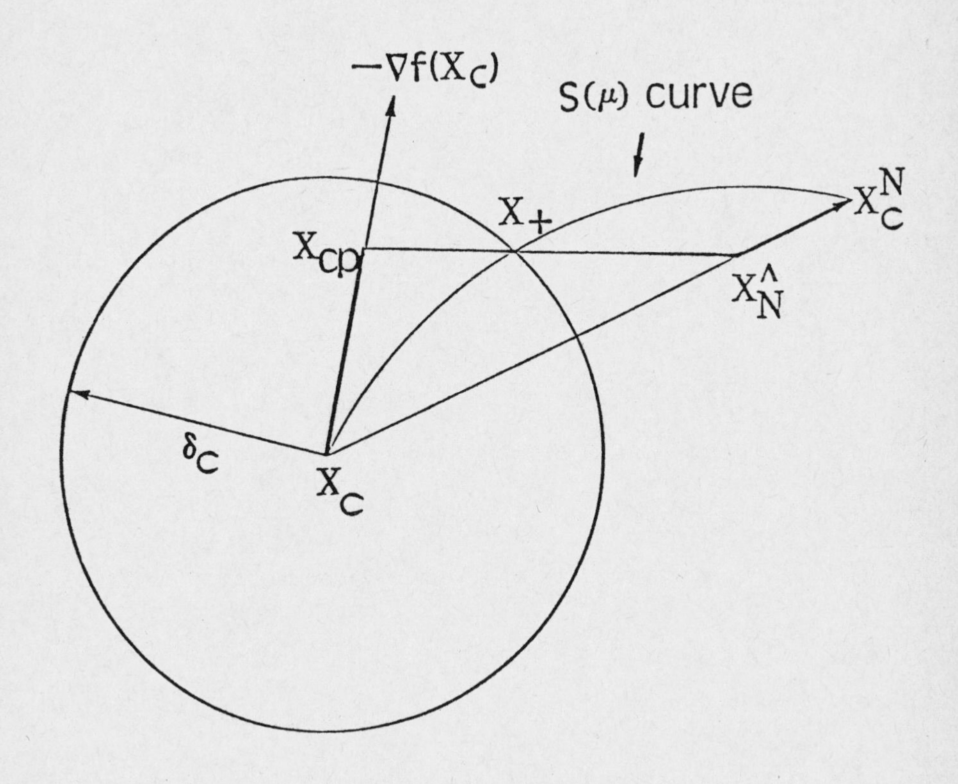

Figure 1. Geometry of the Double Dogleg Step: X_c to X_{cp} to $X_{\hat{N}}$ to X_c^N .

The Cauchy point is the minimizer of the quadratic model in the steepest descent direction (which is the same for both $f(X)$ and $m_c(X)$) and is given by

$$X_{cp} = X_c + S_{cp} \tag{5}$$

where $$S_{cp} = -\lambda \nabla f(X_c),$$

and $$\lambda = \frac{||\nabla f(X_c)||^2}{\nabla f(X_c)^t H_c \nabla f(X_c)} .$$

If $\delta_c < ||S_{cp}||$, then X_{cp} is taken instead as

$$X_{cp} = X_c + \delta_c \, S_{cp} / ||S_{cp}|| \quad . \tag{6}$$

The point in the Newton direction on the double dogleg arc is

$$\hat{X}_N = X_c + \eta S_c^N \quad , \tag{7}$$

where (as recommended by Dennis and Schnabel [6])

$$\eta = 0.8\nu + 0.2 \quad ,$$

ν satisfies

$$||S_{cp}|| \leq \nu ||S_c^N|| \leq ||S_c^N|| \quad ,$$

and is defined by

$$\nu = \frac{||\nabla f(X_c)||^4}{(\nabla f(X_c)^t H_c \nabla f(X_c))(\nabla f(X_c)^t H_c^{-1} f(X_c))} \quad .$$

Then X_+ is the point along the line joining X_{cp} and $\hat{X}_N$ such that $||X_+ - X_c|| = \delta_c$, i.e.

$$X_+ = X_c + S_{cp} + \theta(\hat{X}_N - X_{cp}) \quad , \tag{8}$$

where θ is chosen such that

$$||S_{cp} + \theta(\hat{X}_N - X_{cp})|| = \delta_c \quad .$$

To ensure that the algorithm is making good progress, not only $f(X_+) < f(X_c)$ must be satisfied, but also the criterion

$$f(X_+) \leq f(X_c) + \alpha \nabla f(X_c)^t (X_+ - X_c) \tag{9}$$

(with $\alpha = 10^{-4}$) must be satisfied to guarantee that successive improvements in the function value do not become arbitrarily small. If this condition is not satisfied, then the model function $m_c(X)$ is not representing the true function $f(X)$ well, and the trust region must be reduced. The reduction factor is determined by a backtracking strategy utilizing $f(X_c)$, $f(X_+)$, and the directional derivative $\nabla f(X_c)^t (X_+ - X_c)$ to find a parabola interpolating these data. The new trust region radius δ_{new} corresponds to the minimum of this parabola, and is given by

$$\delta_{new} = - \delta_c \frac{\nabla f(X_c)^t (X_+ - X_c)}{2[f(X_+) - f(X_c) - \nabla f(X_c)^t (X_+ - X_c)]} \quad . \tag{10}$$

The entire double dogleg step and trust region radius calculation (10) are repeated until the acceptability criterion (9) is satisfied.

If X_+ is acceptable, a check is made to see how well m_c has predicted changes in the function f. If the prediction is good, extend the trust region, otherwise, reduce the trust region. The new trust region radius δ_+ is determined as follows:

$$\text{if } |\Delta f| \geqslant 0.75\ |\Delta f_{pred}|\ , \text{ set } \delta_+ = 2\delta_c\ ;$$

$$\text{if } |\Delta f| \geqslant 0.1\ |\Delta f_{pred}|\ , \text{ set } \delta_+ = 1/2\delta_c\ ;$$

$$\text{otherwise leave } \delta_+ = \delta_c\ ,$$

where

$$\Delta f_{pred} = m_c(X_+) - f(X_c)\ ,$$

$$\Delta f = f(X_+) - f(X_c)\ .$$

The double dogleg strategy requires the Jacobian matrix of the system of nonlinear equations which may be singular or ill-conditioned at each iteration. If the Jacobian matrix J is ill-conditioned, the direction $-J^{-1}(X)F(X)$ would be misleading, and the whole strategy would fall apart. The next section discusses a method for dealing with ill-conditioned Jacobian matrices.

Gay's Modified Pseudoinverse Algorithm

Gay [7] shows that if a certain nondegeneracy assumption holds, a modified Newton iteration will converge to a solution of the system of nonlinear equations $F(X)=0$, whose Jacobian matrix exists and is continuous but may be singular at solutions. Consider the Newton iteration

$$X_{k+1} = X_k - J(X_k)^{-1}F(X_k)\ . \tag{11}$$

For $J(X_k)$ singular, $J^{-1}(X_k)$ is undefined, and for $J(X_k)$ nearly singular, a straightforward numerical implementation of (11) encounters serious difficulties. Gay [7] suggests replacing the inverse of the Jacobian matrix in $J^{-1}(X_k)F(X_k)$ by a modified pseudoinverse $\hat{J}(X_k)^+$ in order to maintain numerical stability. Let $J(X_k)$ have the singular value decomposition

$$J(X_k) = USV^t\ ,$$

where U and V are orthogonal matrices, and S is a diagonal matrix with diagonal elements $\sigma_1 \geqslant \sigma_2 \geqslant \ldots \geqslant \sigma_n \geqslant 0$. The Moore-Penrose pseudoinverse of $J(X_k)$ is given by

$$J^+(X_k) = VS^+U^t\ , \tag{12}$$

where S^+ is a diagonal matrix with diagonal elements σ_i^+ given by

$$\sigma_i^+ = \begin{cases} 1/\sigma_i & \text{if} \quad \sigma_i > 0 \\ 0 & \text{if} \quad \sigma_i = 0 \quad . \end{cases}$$

For the Newton iteration using $J^+(X_k)$ in place of $J^{-1}(X_k)$ to be well-defined and convergent, $J^+(X_k)$ has to be continuous and bounded. However, $J^+(X_k)$ is discontinuous at -- and unbounded near -- points X_k where $J(X_k)$ changes rank. Modifications have to be made to produce a continuous substitute $\hat{J}(X)^+$ for $J^+(X)$.

Let $A \in R^{n \times n}$ have singular value decomposition USV^t, $S = \mathrm{diag}(\sigma_1 \ldots, \sigma_n)$. Denote the modified singular values by $\hat{\sigma}_i$, and define

$$\hat{S} = \mathrm{diag}(\hat{\sigma}_1, \ldots, \hat{\sigma}_n) \quad ,$$
$$\hat{A} = U\hat{S}V^t \quad .$$

Fix some tolerance $\varepsilon > 0$. The modified singular values $\hat{\sigma}_i$ are to be chosen such that the following conditions hold for all $\delta > 0$, any matrix $A' = U'S'V'^t$, $\|A - A'\| \leq \delta$, and all $j, k, 1 \leq j, k \leq n$:

$$0 \leq \hat{\sigma}_j^+ \leq 1/\varepsilon \quad , \qquad (13.1)$$

$$|\hat{\sigma}_j^+ - \hat{\sigma}_k'^+| = 0(\delta + |\sigma_j - \sigma_k'|) \quad , \qquad (13.2)$$

$$\sigma_j = \sigma_k \Rightarrow \hat{\sigma}_j^+ = \hat{\sigma}_k^+ \quad , \qquad (13.3)$$

$$\hat{\sigma}_j^+ = 0\,(\sigma_j) \quad , \qquad (13.4)$$

$$0 \leq \sigma_j \hat{\sigma}_j^+ \leq 1 \quad . \qquad (13.5)$$

Under these conditions the modified pseudoinverse

$$\hat{A}^+ = V\hat{S}^+U^t \qquad (14)$$

of A is continuous and bounded in a neighborhood of A where

$$\hat{S}^+ = \mathrm{diag}(\hat{\sigma}_1^+, \ldots, \hat{\sigma}_n^+) \quad ,$$

and

$$\hat{\sigma}_i^+ = \begin{cases} 1/\hat{\sigma}_i & , \quad \hat{\sigma}_i > 0 \\ 0 & , \quad \hat{\sigma}_i = 0 \quad . \end{cases} \qquad (15)$$

A technical requirement for the local convergence of the Newton iteration

$$X_{k+1} = X_k - \hat{J}^+(X_k)F(X_k) \tag{16}$$

to a zero of F(X) is that

$$F(X)^t J(X)\hat{J}(X)^+ F(X) \geq \theta \, \|F(X)\|^2 \tag{17}$$

for some fixed $\theta > 0$ and all relevant $X \in R^n$, where $\hat{J}(X)^+$ is defined by (13-15). Gay [7] has proved that under the nondegeneracy assumption (17) on a C^1 function F(X), the modified Newton iteration (16), where $\hat{J}(X_k)^+$ is the pseudoinverse of $J(X_k)$ modified according to (13-15), converges locally to a zero X_* of F(X), whether or not the Jacobian matrix $J(X_*)$ is singular. The requirement (17) is roughly that (2) has no nonzero local minima. Note that even though Gay's modification provides a numerically stable algorithm in the vicinity of critical points, it may fail if (2) has nearby local minima. The modification also destroys the sparsity of $J(X_k)$, and is only locally convergent in the form of equation (16).

Possible choices of $\hat{\sigma}^+$ satisfying the conditions (13) are

$$\hat{\sigma}^+ = \min[\sigma/\varepsilon^2 \, , \, 1/\sigma] \quad ,$$

$$\text{or} \quad \hat{\sigma}^+ = \sigma/[\sigma^2 + \varepsilon^2/4] \quad ,$$

$$\text{or} \quad \hat{\sigma}^+ = \sigma/[\sigma^2 + \max[0, \varepsilon^2 - \sigma_n^2]] \quad ,$$

where σ_n is the minimum of the singular values.

The numerical results show that Gay's modified pseudoinverse does indeed handle the numerical instability near critical points, and when used judiciously as part of a model trust region strategy permits accurate calculation of equilibrium solutions at and near critical points.

Tunnelling

When the globally convergent quasi-Newton method converges to a local minimum, tunnelling [8-12] is applied to tunnel under all irrelevant local minima, and the method approaches equilibrium solutions in an orderly fashion.

The tunnelling algorithm is designed to achieve a "generalized descent property", that is, to find sequentially local minima of f(X) at X_i^*, i=1,2,...,G, such that

$$f(X_i^*) \geq f(X_{i+1}^*) \quad , \quad i=1,2,\ldots,G-1 \quad , \tag{18}$$

until $f(X)=0$, thus avoiding all local minima that have functional values higher than $f(X_i^*)$.

The tunnelling algorithm is composed of a sequence of cycles. Each cycle consists of two phases, a minimization phase, and a tunnelling phase. The algorithm starts with the minimization phase to find a local minimum. If the local minimum is not an equilibrium solution, the tunnelling phase is entered to obtain a good starting point for minimization in the next cycle. The process is carried on until an equilibrium solution is located.

In the tunnelling phase, the local minimum point X^* is used as a pole. A root X_0 of the tunnelling function

$$T(X,\Gamma) = \frac{f(X)-f(X^*)}{[(X-X^*)^t(X-X^*)]^\lambda} \tag{19}$$

is sought. Γ denotes the set of parameters $(X^*, f(X^*), \lambda)$, where X^* is the current local minimum point, $f(X^*)$ is the functional value of the current local minimum, and λ is the pole strength at X^*. Starting with $\lambda=1$, the pole strength λ at X^* is increased by 0.1 until $T(X,\Gamma)$ decreases away from X^*.

The tunnelling function $T(X,\Gamma)$ itself may have a lot of relative local minima where its gradient is zero. A stabilized Newton method (e.g., model trust region quasi-Newton algorithm) is used in the tunnelling phase to find a X_r such that the gradient $T_x(X_r,\Gamma)$ is equal to zero. If the method converges to a singular point X_m, a movable pole with a pole strength of η is introduced at X_m to cancel the singularity. The tunnelling function becomes

$$T(X,\Gamma) = \frac{f(X)-f(X^*)}{\{[(X-X^*)^t(X-X^*)]^\lambda[(X-X_m)^t(X-X_m)]^\eta\}} \quad . \tag{20}$$

The tunnelling phase stops when $T(X,\Gamma)\leq 0$. Otherwise, X_m is moved to the most recently found relative local minimum X_1. Starting with $\eta=0$, the pole strength η of the movable pole is increased (in increments of 0.1) to enforce a descent property on the nonlinear least squares function of the system $T_x(X,\Gamma)$.

The tunnelling phase is continued until a point X_0 such that $T(X_0,\Gamma)$ is not a local minimum and $T(X_0,\Gamma)\leqslant 0$ is found. Then X_0 is used as the starting point for the next minimization phase. If there is no $X_0 \neq X^*$ such that $f(X_0)\leqslant f(X^*)$, and $f(X^*)\neq 0$, then there is no equilibrium solution at that given load level.

Deflation, as a special case of tunnelling, looks for multiple solutions at a given load level. If X_0^* is an equilibrium solution, another equilibrium solution can be found by locating a zero of the tunnelling function

$$T(X,\Gamma)=[f(X)-f(X_0^*)]/[(X-X_0^*)^t(X-X_0^*)]^\lambda \quad .$$

With the pole strength λ set to 1, the tunnelling function is the same as the deflated function

$$f^*(X)=f(X)/[(X-X_0^*)^t(X-X_0^*)] \quad , \tag{21}$$

since $f(X_0^*)$ is zero. The deflated function is minimized with an initial guess $X_1=(1+\xi)X_0^*$ where ξ is a given perturbation. If a second equilibrium solution X_1^* does exist, the nonlinear least squares function $f(X)$ is deflated with both X_0^* and X_1^*. The deflated function

$$f^*(X)=f(X)/[(X-X_0^*)^t(X-X_0^*)][(X-X_1^*)^t(X-X_1^*)]$$

is minimized to see if a third equilibrium solution exists. The deflation process is continued until no more equilibrium solutions are found.

Some details on minimizing the deflated function follow [13]: In minimizing $f^*(X)$ using the double dogleg strategy, the directions $J^{*t}(X)F^*(X)$ and $-J^{*-1}(X)F^*(X)$ are required, where $J^*(X)$ is the Jacobian of $F^*(X)$. We assumed that minimizing $f^*(X)$ is equivalent to solving the system of nonlinear equations

$$F^*(X) = 0 \quad ,$$

where $f^*(X)=1/2F^{*t}(X)F^*(X)$.

For the single deflation case,

$$F^*(X) = \frac{F(X)}{\|X-X_0^*\|} \quad , \tag{23}$$

for which the Jacobian matrix of $F^*(X)$ is given by

$$J^*(X) = \frac{J(X) + uv^t}{||X-X_0^*||} , \tag{24}$$

where

$$u=-F^*(X) , \quad \text{and} \quad v = \frac{(X-X_0^*)}{||X-X_0^*||} .$$

$J^{*-1}(X)$ can be obtained explicitly as:

$$||X-X_0^*||[J^{-1}(X)-1/\beta J^{-1}(X)uv^tJ^{-1}(X)] , \tag{25}$$

where

$$\beta = 1-v^tp , \quad \text{and} \quad p = J^{-1}(X)F^*(X).$$

Thus

$$-J^{*-1}(X)F^*(X) = -||X-X_0^*||(1/\beta) p , \tag{26}$$

and

$$J^{*t}(X)F^*(X) = \frac{J(X)F^*(X)+vu^tF^*(X)}{||X-X_0^*||} . \tag{27}$$

Similarly, for the double deflation case,

$$-J^{*-1}(X)F^*(X)= -||X-X_0^*|| \; ||X-X_1^*||(1/\beta) p , \tag{28}$$

and

$$J^{*t}(X)F^*(X) = \frac{J^t(X)F^*(X)+vu^tF^*(X)}{||X-X_0^*|| \; ||X-X_1^*||} , \tag{29}$$

with $u = -F^*(X) = \dfrac{-F(X)}{||X-X_0^*|| \; ||X-X_1^*||}$,

$$v = \frac{||X-X_0^*||^2(X-X_1^*) + ||X-X_1^*||^2(X-X_0^*)}{||X-X_0^*|| \; ||X-X_1^*||}$$

$$\beta = 1 - v^tp ,$$

and

$$p = J^{-1}(X)F^{*}(X) \quad .$$

With the above formulas, the quasi-Newton method with the double dogleg strategy can be implemented with deflation using a modified Jacobian matrix while continuing to exploit sparsity and symmetry.

<u>The Overall Algorithm</u>

To solve the system of nonlinear equations

$$F(X_*) = 0 \ ,$$

where $F:R^n \rightarrow R^n$, and $X_* \epsilon R^n$, the following algorithm is applied to minimize the nonlinear least squares function

$$f(X) = 1/2 \ F^t(X)F(X).$$

(1) Start out with an initial tolerance (TOL), an initial guess (X_c), an initial trust region radius (δ_c), and a maximum number of function/Jacobian evaluations (IEVAL).

(2) Calculate the Jacobian matrix $J(X_c)$ from F(X). If the algorithm is minimizing the deflated function, use the modified Jacobian matrix instead.

(3) If the number of function/Jacobian evaluations exceeds IEVAL or $||\nabla f(X_c)|| < TOL$, go to step (12).

(4) Calculate the condition number of the Jacobian matrix. If the Jacobian matrix is ill-conditioned, Gay's modification is applied to perturb the Jacobian matrix into a better conditioned one.

(5) Build a quadratic model m_c around the current estimate X_c.

(6) Calculate the next step $S=S(\mu)=-(H+\mu I)^{-1}\nabla f(X_c)$ such that $||S(\mu)|| \leqslant \delta_c$ by the double dogleg strategy to minimize the nonlinear least squares function f(X).

(7) Calculate $F(X_c+S)$. If the number of function/Jacobian evaluations exceeds IEVAL, go to step (12). If the step S is acceptable, go to step (8). Otherwise go to step (9).

(8) The step S is acceptable: Set $X_+ := X_c+S$. If $S=S(0)$ (the trust region includes the quasi-Newton point), go to step (11). Otherwise go to step (10).

(9) The step S is not acceptable. If the algorithm is not trying to take a bigger step, then reduce the trust region radius by a factor determined from a backtracking strategy, and go back to step (6). Otherwise, restore the good X_+ and $f(X_+)$ that was saved in step (10) before, and to go step (11).

(10) If the actual reduction and the predicted reduction are in good agreement or the reduction in the true function is large, then save the X_+ and $f(X_+)$. Go back to step (6) to try a bigger step by doubling the trust region radius.

(11) If $||F(X_+)|| > TOL$, update the trust region according to the prediction of the function f(X) by the model function $m_c(X)$. Set $X_c := X_+$, and go back to step (2). Otherwise, go to step (13).

(12) $X_0 = X_c$ is a local minimum or the algorithm has failed to make significant progress. Tunnelling is applied to find a X_0^+ such that $T(X_0^+, \Gamma) < 0$. If such a X_0^+ exists, reset IEVAL:=0, reset the initial guess $X_c := X_0^+$, and go back to step (2). Otherwise, $f(X_c)$ is the global minimum, and the algorithm stops since there are no more equilibrium solutions at the given load level.

(13) $X^* = X_+$ is an equilibrium solution. If more equilibrium solutions at that load level are desired, deflate the nonlinear least squares function with the solution X^*, reset IEVAL:=0, reset the initial guess $X_c := (1+\xi)X^*$, and go back to step (2). Otherwise, the algorithm stops.

In the implementation of the proposed method, two poles, one at the most recently found local minimum (X_{1m}), and the other at the most recent found equilibrium solution (X_{es}), are introduced in the nonlinear least squares function f(X) to form the deflated function $f^*(X)$ that is minimized. If the method converges to a new local minimum or equilibrium solution, the corresponding pole is moved to that local minimum or equilibrium solution. The process is carried on until either the desired number of equilibrium solutions is found, there are no more equilibrium solutions (see step (12)), or a limit on the number of function/Jacobian evaluations is exceeded. In the tunnelling phase, instead of a stabilized Newton Method, a quasi-Newton method with the double dogleg strategy is used. The quasi-Newton method finds an X_0 such that the tunnelling function $T(X,\Gamma)$ is less than zero, then X_0 is used as the starting point in the next minimization phase, and the algorithm proceeds.

Illustration of the Proposed Method

The proposed method was first validated on the snap-through response of a shallow arch shown in Figure 2. The load deflection curve of the crown of the arch has two limit points, one at 1773.00 lb, and the other at 3064.18 lb. The task was to locate all equilibrium solutions at every load level. The load deflection curve of the crown of the arch was found by tracking the curve with the homotopy method of Kamat, Watson and Venkayya [5]. At each load level, the equilibrium solutions that were located by the proposed method are denoted by a '*'. Figure 3 shows that all equilibrium solutions at each level were located successfully by the proposed method.

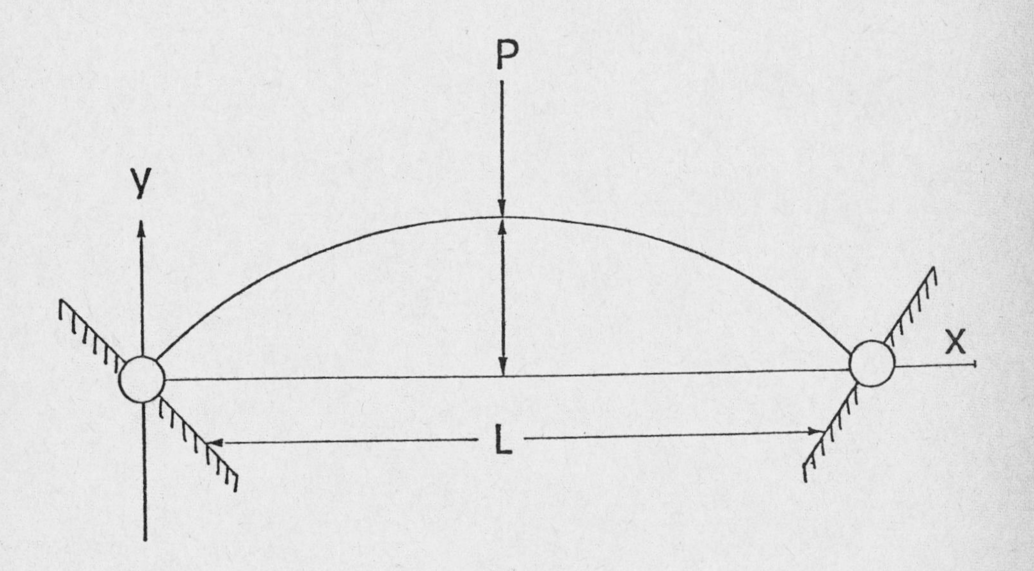

Figure 2. Shallow Arch (29 degrees of freedom):
$y = a \sin(bx/L)$,
$a = 5$ in.,
$L = 100$ in.,
$A = 0.32$ in^2.,
$I = 1$ in^4.,
$E = 10^7$psi,
10 frame elements for 1/2 span.

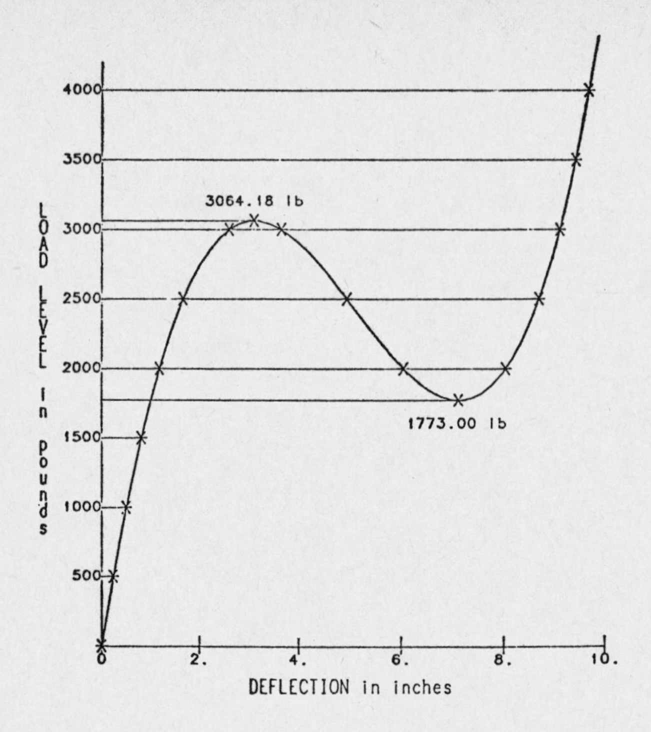

Figure 3. Load Deflection Curve of Shallow Arch Crown.

After the first equilibrium solution was found, deflation was applied to locate the second and third equilibrium solutions (if they exist). The method was initiated with a load of 500 lb, and an increment of 500 lb for the next load steps. The first equilibrium solution of a given load step was used as an initial guess for the next load step. As shown in Figure 3, the first three load steps had only one equilibrium solution, the fourth to sixth load steps had three, and there was only one equilibrium solution each for the seventh and eighth load steps. At the seventh load step (3500 lb), when the first equilibrium solution of the sixth load step was used as the initial guess, the proposed method converged to a local minimum. Tunnelling was applied, and the distant equilibrium solution was located. When the load steps were close to limit loads, Gay's modification was applied to perturb the Jacobian matrix into a better conditioned one to accelerate convergence to the equilibrium solution.

The proposed method was compared to a classical Newton method and a quasi-Newton method using the double dogleg strategy (QNM-DDL) but without Gay's modification in the vicinity of limit points. The classical Newton method diverged, while QNM-DDL without Gay's modification failed to locate the equilibrium solutions near limit points. Both methods failed in the vicinity of limit points due to the ill-conditioning of the Jacobian matrix. For larger load steps the classical Newton method diverged.

The proposed method checks the condition number of the Jacobian matrix at every iteration. If it is necessary to locate equilibrium solutions in the vicinity of critical points, Gay's modification is invoked to perturb the Jacobian matrix into a better conditioned one, since the Jacobian matrix is ill-conditioned near critical points. Otherwise, it uses a standard quasi-Newton method with the double dogleg strategy to locate a minimum of the least squares function (2).

The method was also validated on the snap-through response of a shallow dome under a concentrated load at the center, as shown in Figure 4. The load response curve of the crown of the shallow dome is quite complex, having many equilibrium solutions at each load level. Since there are many bifurcation points, only a portion of the load response curve is shown in Figure 5. The equilibrium solutions that were located by the proposed method are indicated by a '*' in Figure 5. As shown in Figure 5, multiple equilibrium solutions at each load level were located to illustrate the success of the proposed method.

The proposed method only deflates using at most two poles: one at the most recently found local minimum point, and the other at the most recently found equilibrium solution (if they exist). However, since the proposed method may converge back to a recently found equilibrium solution or local minimum, and the application of deflation may virtually destroy some nearby minima, it is not guaranteed that all equilibrium solutions can be located. Of course, we could deflate with more than two poles, but then the algorithm quickly becomes unwieldy, and there is still no guarantee of success. Here we only try to illustrate the success of the proposed method in finding multiple (but not necessarily all) equilibrium solutions.

In applying the tunnelling algorithm, the initial guess in the tunnelling phase has to be far away from the current local minimum point to prevent getting back this same local minimum point as the next starting point. Since the local minimum point that was found is only an approximation to the true local minimum point, if we start the

Coordinates of the Node Points of Dome Structure.

Node	X	Y	Z
1	0.0	0.0	6.0
3	-15.0	25.9807	4.5
4	-30.0	0.0	4.5
9	0.0	60.0	0.0
10	-30.0	51.9615	0.0
11	-51.9615	30.0	0.0
12	-60.0	0.0	0.0

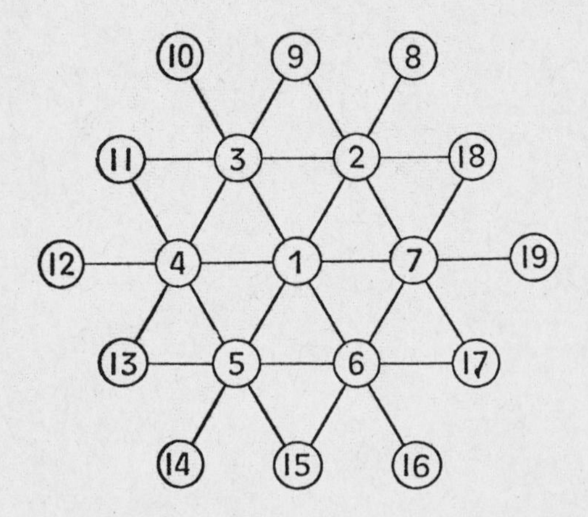

Figure 4. Shallow Dome (21 degrees of freedom):

$A_i = 0.1$ in^2., $i = 1, 2, \ldots, 30$,

nodes 1 to 7 are free,

nodes 8 to 19 are pinned.

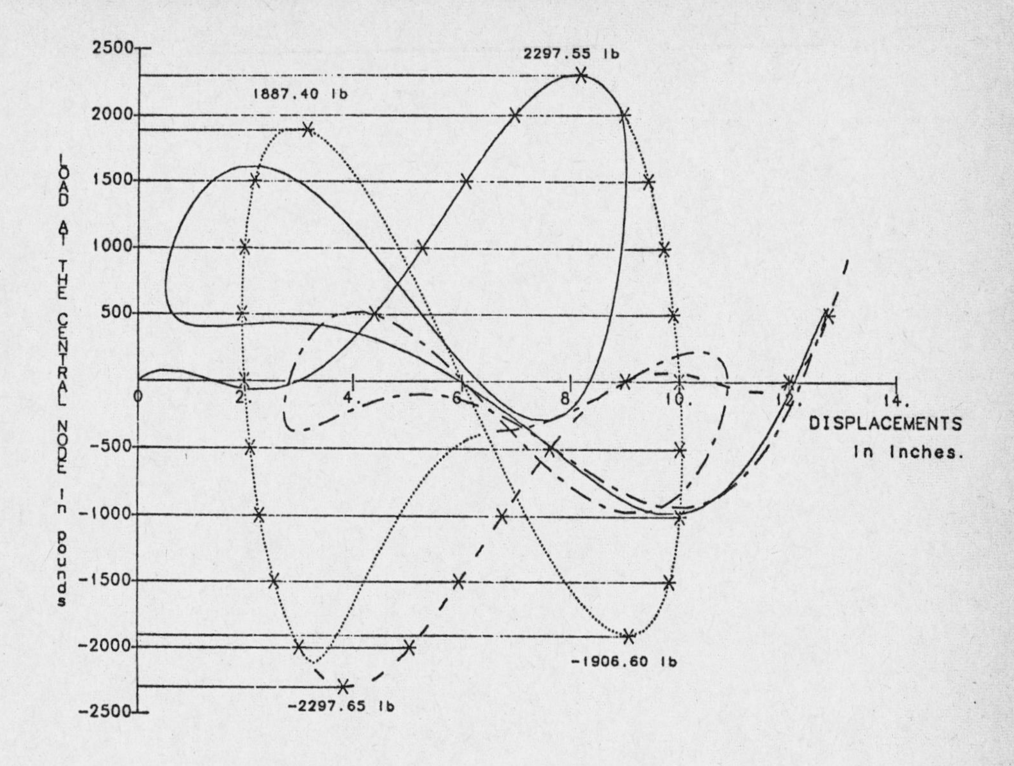

Figure 5. Load Deflection Curve of Shallow Dome Crown.

tunnelling phase with an initial guess close to the (computed) local minimum point, it may converge to another approximation of the <u>same</u> local minimum point but with a lower functional value. In this case the minimization phase and the next tunnelling phase produce again the same approximate local minimum point, and no progress is being made. However, if the initial guess is far away from the current local minimum point, there is a chance that some equilibrium solutions will be missed.

The curve in Figure 5 was generated by starting the homotopy method of Kamat, Waton and Venkayya [5] with different starting points and an accuracy of 10^{-10}. Due to the complicated response of the structure to the loads, there are many bifurcation points along the curve, which the homotopy method is not designed to handle. Neither the homotopy nor the quasi-Newton algorithm by itself could have produced all the branches shown in Figure 5. The curve in Figure 5 was generated by starting the homotopy method at zero first (the solid line), and then from the equilibrium solutions that were located by the proposed model trust region quasi-Newton method (the dashed, dotted, and dashed-dotted lines). For simplicity only portions of the equilibrium curves are shown in Figure 5. Some points on the curves that were not located by the quasi-Newton method were validated by starting the quasi-Newton method nearby, and the quasi-Newton method converged to the same equilibrium solutions computed by the homotopy method.

Conclusion

The proposed method, model trust region quasi-Newton algorithm and tunnelling, works extremely well in locating multiple equilibrium solutions, either stable or unstable. Although the use of Gay's modification in the vicinity of critical points destroys sparsity and symmetry, it is only invoked when equilibrium solutions in the vicinity of critical points are needed.

As an alternative to the hybrid method of Kamat and Watson [13], the model trust region quasi-Newton method with tunnelling is a more efficient method for locating a few equilibrium solutions. With the use of deflation, multiple solutions can be located. If equilibrium solutions in the vicinity of critical points are desired, they can be located by using Gay's modified pseudoinverse. With the use of a skyline structure to store the Jacobian matrix, the method exploits sparsity and symmetry. Update formulas for the Jacobian matrix, such as Toint's update [15], can be used in the future to save the costly Jacobian evaluations at every iteration. On the other hand, the proposed algorithm may present a fragmented picture. For example, Figure 5 would have been very difficult to obtain without the homotopy method. Although more has to be done in the future to make the method robust, the preliminary results are promising.

REFERENCES

[1] G.A. Wempner, Discrete Approximations Related to Nonlinear Theories of Solids, International Journal of Solids and Structures 7 (1971) 1581-1599.

[2] E. Riks, An Incremental Approach to the Solution of Snapping and Buckling Problems, International Journal of Solids and Structures 15 (1979) 529-551.

[3] M.A. Crisfield, A Fast Incremental/Iterative Solution Procedure that Handles "Snap-Through", Computers and Structures 13 (1981) 55-62.

[4] J. Padovan, Self-Adaptive Predictor-Corrector Algorithm for Static Nonlinear Structural Analysis, NASA CR-165410 (1981).

[5] M.P. Kamat, L.T. Watson, and V.B. Venkayya, A Quasi-Newton versus a Homotopy Method for Nonlinear Structural Analysis, Computers and Structures 17, No. 4 (1983) 579-585.

[6] J.E. Dennis Jr. and R.B. Schnabel, Numerical Methods for Unconstrained Optimization and Nonlinear Equations, Prentice-Hall, Englewood Cliffs, N.J. (1983).

[7] D. Gay, Modifying Singular Values: Existence of Solutions to Systems of Nonlinear Equations Having a Possible Singular Jacobian Matrix, Mathematics of Computation 31 (1977) 962-973.

[8] S. Gómez and A.V. Levy, The Tunnelling Algorithm for the Global Optimization of Constrained Functions, Comunicaciones Técnicas (1980), Serie Naranja No. 231, IIMAS-UNAM.

[9] A.V. Levy and A. Montalvo, The Tunnelling Algorithm for the Global Minimization of Functions, Dundee Biennal Conference on Numerical Analysis (1977), University of Dundee, Scotland.

[10] A.V. Levy and A. Montalvo, Algoritmo de Tunelización para la Optimación Global de Funciones, Comunicaciones Técnicas (1979), Serie Naranja No. 204, IIMAS-UNAM.

[11] A.V. Levy and A. Calderón, A Robust Algorithm for Solving Systems of Non-linear Equations, Dundee Biennal Conference on Numerical Analysis (1979), University of Dundee, Scotland.

[12] A.V. Levy and A. Montalvo, A Modification to the Tunnelling Algorithm for Finding the Global Minima of an Arbitrary One Dimensional Function, Comunicaciones Técnicas (1980), Serie Naranja No. 240, IIMAS-UNAM.

[13] M.P. Kamat, L.T. Watson, and J.L. Junkins, A Robust and Efficient Hybrid Method for Finding Multiple Equilibrium Solutions, Proc. Third International Symposium on Numerical Methods in Engineering, Paris, France, Vol. II, (1983) 799-807.

[14] Ph.L. Toint, On Sparse and Symmetric Matrix Updating Subject to a Linear Equation, Math. Comp. 31 (1977) 954-961.

[15] J.E. Dennis Jr. and J.J. Moré, Quasi-Newton Methods: Motivation and Theory, SIAM Review 19 (1979) 46-89.

[16] J.E. Dennis Jr. and R.B. Schnabel, Least Change Secant Updates for Quasi-Newton Methods, SIAM Review 21 (1979) 443-459.

CONSIDERATIONS OF NUMERICAL ANALYSIS IN A SEQUENTIAL QUADRATIC PROGRAMMING METHOD

by

Philip E. Gill, Walter Murray,
Michael A. Saunders and Margaret H. Wright

Systems Optimization Laboratory
Department of Operations Research
Stanford University
Stanford, California 94305, USA

ABSTRACT

This paper describes some of the important issues of numerical analysis in implementing a sequential quadratic programming method for nonlinearly constrained optimization. We consider the separate treatment of linear constraints, design of a specialized quadratic programming algorithm, and control of ill-conditioning. The results of applying the method to two specific examples are analyzed in detail.

1. Overview of a sequential quadratic programming method

The general *nonlinear programming problem* involves minimization of a nonlinear objective function subject to a set of nonlinear constraints. Sequential quadratic programming (SQP) methods are widely considered today as the most effective general techniques for solving nonlinear programs. The idea of treating constraint nonlinearities by formulating a sequence of quadratic programming subproblems based on the Lagrangian function was first suggested by Wilson (1963). A brief history of SQP methods and an extensive bibliography are given in Gill, Murray and Wright (1981). Powell (1983) gives a survey of recent results and references.

SQP methods have been (and remain) the subject of much research, particularly concerning theoretical properties such as global and superlinear convergence. However, the enormous recent interest in SQP methods has arisen primarily because of their remarkable success in practice. Therefore, this paper is devoted to selected issues of *numerical analysis* that have arisen in the implementation of a particular SQP method (the code NPSOL; see Gill *et al.*, 1984a). We have restricted our attention to a single method in order to analyze the issues in detail. However, the same ideas can be applied to SQP methods in general.

We assume that the problem to be solved is of the form

$$\text{NP} \qquad \begin{array}{ll} \underset{x\in\Re^n}{\text{minimize}} & F(x) \\ \text{subject to} & \ell \le \left\{ \begin{array}{c} x \\ \mathcal{A}_L x \\ c(x) \end{array} \right\} \le u, \end{array}$$

where $F(x)$ is a smooth nonlinear function, $\mathcal{A}_L$ is a constant matrix with m_L rows, and $c(x)$ is a vector of m_N smooth nonlinear constraint functions. (Both m_L and m_N may be zero.) The i-th constraint is treated as an *equality* if $\ell_i = u_i$; components of ℓ or u can be set to special values that will be treated as $-\infty$ or $+\infty$ if any bound is not present.

Let x_k denote the k-th estimate of the solution of NP. The next iterate is obtained during the k-th *major iteration*, and is defined by

$$x_{k+1} = x_k + \alpha_k p_k. \tag{1}$$

The n-vector p_k (the *search direction*) is the solution of a quadratic programming (QP) subproblem. In order to specify a QP subproblem, we must define a set of linear constraints and a quadratic objective function. The linear constraints of the subproblem are linearizations of the problem constraints about the current point. The quadratic objective function of the subproblem approximates (in some sense) the *Lagrangian function*, which is a special combination of the objective and constraint functions. The curvature of the Lagrangian function plays a critical role in the optimality conditions for NP (for further discussion, see, e.g., Powell, 1974; Gill, Murray and Wright, 1981). Thus, the subproblem has the following form:

$$\begin{array}{ll} \underset{p\in\Re^n}{\text{minimize}} & g^T p + \frac{1}{2} p^T H p \\ \text{subject to} & \bar{\ell} \le \left\{ \begin{array}{c} p \\ \mathcal{A}_L p \\ \mathcal{A}_N p \end{array} \right\} \le \bar{u}, \end{array} \tag{2}$$

where the vector g is the gradient of F at x_k, the symmetric matrix H is a positive-definite quasi-Newton approximation to the Hessian of the Lagrangian function, and $\mathcal{A}_N$ is the Jacobian matrix of c evaluated at x_k. The vectors $\bar{\ell}$ and $\bar{u}$ contain the constraint residuals at x_k with respect to the original bounds.

The non-negative *step length* α_k in (1) is chosen so that x_{k+1} exhibits a "sufficient decrease" (Ortega and Rheinboldt, 1970) in a *merit function*, which is intended to ensure progress toward the solution by balancing improvements in the objective function and constraint violations. The merit function in our algorithm is a smooth augmented Lagrangian function (see, e.g., Schittkowski, 1981, 1982; Gill *et al.*, 1984a). A notable feature is that nonlinear inequality constraints are treated using simply-bounded slack variables. At each major iteration, the line search is performed with respect to the variables x, multiplier estimates λ, and slack variables s; all these elements are available from the solution of the QP subproblem (2). The steplength is required to produce a sufficient decrease in the merit function

$$\mathcal{L}(x,\lambda,s) = F(x) - \sum_{i=1}^{m} \lambda_i\big(c_i(x) - s_i\big) + \frac{\rho}{2}\sum_{i=1}^{m}\big(c_i(x) - s_i\big)^2.$$

(A detailed description of the definition of the slack variables is given in Gill *et al.*, 1985a.) The value of the *penalty parameter* ρ is initially set to zero, and is occasionally increased from its value in the previous iteration in order to ensure descent for the merit function. The sequence of penalty parameters is generally non-decreasing, although we allow ρ to be reduced a limited number of times.

In the remainder of this paper, we shall concentrate on three topics in numerical analysis that affect the efficiency and reliability of an SQP method: techniques for exploiting constraint linearities; design of a QP method specifically tailored for use within an SQP method; and monitoring and control of ill-conditioning.

2. Treatment of linear constraints

In theoretical discussions of SQP methods, it is customary (and reasonable) to assume that all constraints are nonlinear. However, in developing an implementation of an SQP method, one must ask whether explicit treatment of linear constraints would lead to improvements in the algorithm itself and/or in computational efficiency. It makes sense to consider separate treatment of linear constraints because the problem formulator usually knows about constraint linearities, in contrast to other properties that might be exploited in an SQP method (such as convexity), but that are difficult or impossible to determine. Furthermore, many problems arising in practice — particularly from large-scale models — contain a substantial proportion of linear constraints.

Our choice of NP as the problem formulation implies that linear constraints and bounds on the variables are indeed represented separately from nonlinear constraints. In this section, we consider the reasons for, and implications of, this decision. (The issue of the further distinction between simple bounds and general linear constraints has been discussed in Gill *et al.*, 1984b.)

It is well known that problems with only linear constraints are typically much easier to solve than problems with nonlinear constraints, for several reasons (both algorithmic and practical). First, the optimality conditions for linear constraints are much less complex. Second, *finite* (and efficient) procedures (e.g., a phase-1 simplex method) are known for finding a feasible point with respect to a set of linear constraints, or determining that there is none. In contrast, there is no general guaranteed procedure for computing a feasible point with respect to even a single nonlinear constraint, unless the constraints have certain properties. Third, once an initial feasible point has been found with respect to a set of linear constraints, feasibility can be assured for all subsequent iterations within a method whose iteration is defined by (1) by suitable definition of the search direction and step length. For nonlinear constraints, an iterative procedure would typically be required to restore feasibility. Finally, the gradient of a linear function is constant, and hence needs to be computed only once.

It might be argued that an SQP method "automatically" takes advantage of constraint linearities, since the linearization of a linear constraint is simply the constraint itself. However, treating all constraints uniformly would have certain undesirable implications. The iterates would not necessarily be feasible with respect to the linear constraints until after an iteration in which $\alpha_k = 1$. Since feasibility with respect to linear constraints can be retained in a straightforward manner, their inclusion in the merit function seems unnecessary, particularly since they would appear nonlinearly. Finally, the gradients of linear constraints would be unnecessarily re-evaluated at every major iteration.

Based on these considerations, it was decided to treat linear constraints separately. The obvious next question is: how far should the separate treatment extend? It is interesting that, as successive versions of the algorithm have been developed, the degree to which linear constraints are treated separately has consistently *increased*.

In the present algorithm, linear constraints are treated specially even before the first QP subproblem is posed. In effect, the nonlinear constraints are (temporarily) ignored, and a phase-1 procedure is executed to find a feasible point with respect to the linear constraints. (Thus, the starting point provided by the user is not necessarily the initial point for the nonlinear phase of the algorithm.) Our justification for this approach is two-fold. A "practical" reason is that in many problems it is not uncommon for some or all of the nonlinear functions to be undefined or poorly behaved at points where the linear constraints (particularly the bounds) are not satisfied. The more important algorithmic reason arises from the strategies used by SQP methods to deal with infeasible QP subproblems.

Any robust SQP method must be able to cope with inconsistency in the constraints of the QP subproblem (see, e.g., Powell, 1977a; Schittkowski, 1982; Tone, 1983). Most techniques for doing so involve solving additional subproblems. Such strategies are based on the (optimistic) assumption that the inconsistency is *temporary*, since infeasible linearized constraints at one point do not necessarily imply inconsistency of the original nonlinear constraints. Our motivation for looking first at just the linear constraints is to determine whether the problem is *inherently* infeasible. Obviously, it is useless to solve a sequence of modified subproblems if the original linear constraints are themselves inconsistent. By first ensuring feasibility with respect to the linear constraints, we guarantee that any inconsistency in a subproblem is attributable only to constraint nonlinearities.

It might appear that a separate phase-1 procedure involving only the linear constraints would lead to extra work, since the next QP to be solved includes all the constraints from the phase-1 procedure as well as the linearized nonlinear constraints. However, as we shall see in the next section, advantage can be taken of the information gained during the initial phase-1 procedure.

A decision to retain feasibility with respect to linear constraints has other implications as well, some of which make the implementation more complicated. For example, the step length procedure permits α_k in (1) to assume values greater than unity if a significant reduction in the merit function can thereby be obtained. (This feature is not typically found in other SQP methods.) Therefore, feasibility with respect to the linear constraints can be ensured only if the value of α_k in (1) is bounded above by the step to the nearest linear constraint (as in standard active-set methods).

3. A specialized quadratic programming algorithm

The second issue to be considered is the benefit of designing a QP method intended specifically for use within SQP methods. In the early days of SQP methods, it was believed that any good "off the shelf" algorithm could be used to solve the QP subproblem. (A similar desire to use "black boxes" has been observed since the first production of mathematical software.) However, it is now generally agreed that substantial gains in efficiency can result from a suitably tailored QP algorithm. We emphasize that *the improvements do not result from a decrease in the number of evaluations of the user-provided functions.* Rather, substantial reductions can be achieved in the linear algebra, which may comprise a substantial fraction of the total solution time, even for problems of moderate size.

Development of the QP method to be described was motivated by the special features of the QP subproblems associated with SQP methods. However, the net result has been to *build a better black box*, since the QP method can be used with equal success on general problems.

3.1. Background on active-set QP methods. Most modern quadratic programming methods are *active-set* methods, which essentially involve an iterative search for the correct active set (the subset of constraints that hold with equality at the solution). (The iterations within the QP method itself will be called *minor* iterations.) In our QP algorithm, an initial feasible point is found to serve as the first iterate. At a typical iteration, let C denote a *working set* of m constraints that hold exactly at the current iterate p; let $\bar{g}$ denote the gradient of the quadratic function at p; and let Z denote a basis for the set of vectors orthogonal to C, i.e., such that $CZ = 0$. By solving the well known *augmented system*

$$\begin{pmatrix} H & C^T \\ C & 0 \end{pmatrix} \begin{pmatrix} d \\ -\mu \end{pmatrix} = \begin{pmatrix} -\bar{g} \\ 0 \end{pmatrix}, \tag{3}$$

we obtain d (the step to the minimum of the quadratic function subject to the working set held at equality) and μ (the Lagrange multipliers at $p + d$). If C has full rank and H is non-singular, the solution of (3) is unique. *In order to verify optimality, the system (3) must be solved at least once, even if C is the correct active set.*

In our QP algorithm (the code QPSOL; see Gill *et al.*, 1984c, for details), (3) is solved using the TQ factorization of C:

$$CQ \equiv C(Z \quad Y) = (0 \quad T),$$

where Q is orthogonal and T is reverse-triangular. Note that Z comprises the first $n-m$ columns of Q. We also require the Cholesky factorization of Z^THZ (the *projected Hessian*).

Changes in the working set occur for two reasons. If the full step of unity cannot be taken along d because a constraint not in the working set would thereby be violated, a restricted step is taken and the given constraint is *added* to the working set. If a step of unity can be taken, but some component of μ is negative, the corresponding constraint is deleted from the working set after taking the unit step. Each change in the working set leads to changes in T, Q and Z^THZ. When bounds are treated separately, both rows and columns of C change. (See Gill *et al.*, 1984b, for details of the update procedures.)

3.2. The "warm start" procedure. When solving NP by an SQP method, the active sets of the QP subproblems eventually become the same set of indices as the active set of the original nonlinear problem (see, e.g., Robinson, 1974, for a proof). Thus, at least in the neighborhood of the solution (and sometimes far away), *the active sets of successive QP subproblems will be the same.* Since the essence of a QP method is a search for the active set, it would obviously be highly desirable to exploit *a priori* knowledge of the active set.

To do so, the QP method has been extended to include a "warm start" feature, similar to those in linear programming codes. The basic idea is that the user has the option of specifying a "predicted" working set $\tilde{C}$ as an input to the QP, along with a starting point $\tilde{p}$. The warm start procedure does not merely check whether the specified constraints are satisfied exactly at $\tilde{p}$. Rather, it constructs and factorizes a linearly independent working set C (see Section 4.1) — preferably $\tilde{C}$. It then computes δ, the minimum change to $\tilde{p}$ such that the constraints in C are satisfied exactly at $\tilde{p}+\delta$. The initial point of the phase-1 procedure is taken as $p_0=\tilde{p}+\delta$.

Within the SQP method, the active set of each QP subproblem is used as the predicted initial working set for the next, with $\tilde{p}$ taken as zero. Since the active sets eventually become the same, the effect of the warm start procedure is that later QP subproblems reach optimality in *only one iteration*. We emphasize this point because of an unfortunate misapprehension that the need for a feasibility phase implies that a QP method such as that of Section 3.1 will be *inefficient* in an SQP method. In fact, the opposite is true.

To see why, consider the effect of using a warm start option "sufficiently near" the optimal solution of NP so that $\tilde{C}$ contains linearizations of the correct active set. The initial point p_0 for the feasibility phase will satisfy

$$\tilde{C}p_0=-c,$$

where c is the vector of violations of the active constraints. The constraints inactive at the solution of NP are strictly satisfied in the neighborhood of the solution; thus, the zero vector is feasible with respect to the linearized inactive constraints near the solution. If $\|c\|$ is "small" and $\tilde{C}$ is not ill-conditioned, $\|p_0\|$ will also be "small", and will remain feasible with respect to the linearized inactive constraints. Therefore, the vector produced by the warm start procedure will be feasible with respect to *all* the QP constraints, without any phase-1 iterations.

For reasonably well-behaved problems, the work associated with solving the later QP subproblems with a warm start option is equivalent to solving a single system of equations of the form (3) — the minimum amount of work necessary to solve a QP. It would be impossible to solve the QP more efficiently! The improvements in efficiency from the warm start option depend only on the characteristic of SQP methods that the active sets of successive subproblems are the same from some point onwards. Its effect can be seen by examining the number of minor (QP) iterations associated with each subproblem, and will be illustrated by example in Section 5.

3.3. The "hot start" procedure. Given the gains from adding a warm start procedure to the QP method, it is natural to ask: can we do better? It turns out that the answer is "yes", *if linear constraints are treated separately.*

Consider the formulation of a typical QP subproblem. Since the predicted working set is taken as the active set of the previous QP subproblem, the initial working set of the new subproblem is given by

$$C = \begin{pmatrix} A_L \\ \bar{A}_N \end{pmatrix}. \tag{4}$$

The matrix A_L corresponds to linear constraints whose indices occur first in the working set, and thus remains *constant*. The matrix $\bar{A}_N$ includes the gradients of nonlinear constraints as well as of linear constraints that may have been added to the active set during the solution of the previous QP subproblem. The first step of the QP method is to compute the TQ factorization of (4). *If the matrix Q from the previous QP is available*, this factorization need not be computed from scratch. Since the existing Q already triangularizes A_L, i.e.,

$$CQ = \begin{pmatrix} A_L \\ \bar{A}_N \end{pmatrix} Q = \begin{pmatrix} 0 & T_L \\ S & \bar{W} \end{pmatrix},$$

only the matrix S needs to be triangularized in order to complete the factorization of C. By testing the indices of constraints in the active set, it is possible to determine exactly how much of the previous factorization can be saved.

This feature of the QP algorithm is called a "hot start", and can lead to great savings when the problem contains a significant proportion of linear constraints. Note that a special-purpose QP algorithm is necessary in order to accept the TQ factorization as initial data. In contrast to the warm start option, the hot start feature does not affect the number of QP iterations. Rather, it reduces *the amount of work associated with the first QP iteration.* For later major iterations that involve only a single QP iteration, the hot start option leads to significant gains in speed. Furthermore, since the first QP iteration is always much more expensive than subsequent iterations, a decrease in cost of the first iteration is equivalent to a large reduction in the number of QP iterations.

Further savings can be obtained by developing an even more specialized QP algorithm that utilizes as initial data the Cholesky factorization of $Q^T HQ$ in addition to the TQ factorization of the predicted working set. The justification for such a QP algorithm is our empirical observation that, even with the hot start option, the first QP iteration is sometimes quite expensive because of the need to compute the Cholesky factorization of the projected Hessian from scratch. If A_L consistently constitutes a substantial portion of C, the Cholesky factors of Q^THQ can be *updated* to reflect the changes in Q associated with $\bar{A}_N$.

3.4. Continuity of Z. An interesting (and important) result of the development of the QP code has been to ensure the continuity of the matrix Z associated with each major iteration of the SQP method. In theoretical proofs of convergence for SQP methods that maintain an approximation to the projected Hessian of the Lagrangian function, it is crucial that small changes in x should lead to small changes in Z (see, e.g., Coleman and Conn, 1982, 1984). Coleman and Sorensen (1984) have recently observed that the "standard" way of computing Z — by triangularizing the predicted active set with Householder transformations from scratch at each new point — leads to inherent discontinuities in Z, even in the neighborhood of a point where A has full rank. With the procedure described above, in which Q is obtained by *updating* the old Q, it can be shown that Z is *continuous* in the neighborhood of a point at which A has full rank. Furthermore, the change in Z is uniformly bounded in the neighborhood, and Z converges to a limit if the iterates $\{x_k\}$ converge sufficiently fast to x^*. Continuity is preserved because, if the change in x is "small", the procedure triangularizes an almost-triangular matrix. It can be shown that the associated sequence of Householder transformations makes "small" changes *in the columns of* Z, even though Q itself is not continuous (see Gill *et al.*, 1983). With this procedure, Z depends on the previous Q as well as on x, and the limiting Z depends upon the sequence $\{x_k\}$.

With the standard Householder procedure, all of Q is not continuous because the columns of Y change sign with every update. However, uniform continuity in *all of* Q can be achieved by performing the updates with *regularized* Householder transformations (which differ from the standard ones by a change of sign in one row) (see Gill *et al.*, 1985c).

4. Numerical stability

This section turns to a topic of critical importance in numerical analysis — numerical stability. We have selected a few aspects for detailed discussion, extending from the minor iterations of the QP method to the major iterations.

Both the SQP and QP methods are significantly affected by the conditioning of the active (or working) set, which we shall denote for convenience by C. Ill-conditioning in C leads to inaccurate (usually large) search directions, since the search direction must satisfy the equations $Cp = -c$. Furthermore, the Lagrange multipliers of the QP are computed from other equations that involve C:

$$C^T\mu = g + Hp.$$

Hence, an ill-conditioned C tends to produce large (and inaccurate) Lagrange multipliers, which in turn adversely affect the quasi-Newton update and hence the approximate Hessian of the next QP subproblem.

In order to devise strategies to control the condition of C, we must have a simple and inexpensive way of measuring the condition number of C. Fortunately, a good *estimate* of the condition number of the working set can be obtained from its TQ factorization (3) — the ratio of the largest to the smallest diagonal elements of T (denoted by $\tau(C)$). Although $\tau(C)$ is only a lower bound on the condition number, in practice it provides a good indication of the general trend of the condition of the working set. In the rest of this section, the "condition number" should be interpreted as the *estimated* condition number.

Given an ability to compute τ, is it actually possible to exercise any control over the condition of the working set in the SQP or QP algorithms? (If we are given an ill-conditioned linear system to solve, we cannot simply ignore some of the equations!) Unless the constraints are exactly dependent, the active set of a nonlinear or quadratic programming problem is not subject to control by the algorithm used to solve it. However, one feature of a QP method is that the working set *changes*. Thus, we have some freedom in choosing the working set as the QP iterations proceed so as to maintain the best possible condition estimate, and to avoid *unnecessary* ill-conditioning. In the QP method, the condition of the working set is controlled in two ways: directly and indirectly.

4.1. The initial working set. Recall that the phase-1 procedure constructs an initial working set and then computes its TQ factorization. During this part of the algorithm, *direct* control of the condition number is possible with the following strategy. The phase-1 procedure is initiated by adding the desired bounds to the working set. A working set composed only of bounds is essentially "free", and is perfectly conditioned (it is simply a submatrix of the identity). Furthermore, a bound in the working set corresponds to *removing* the corresponding column in the matrix of general constraints, thereby reducing the dimension of the matrix whose TQ factorization must be computed.

Having added all the desired bounds, the set of candidate general constraints is processed. As general constraints are added to the working set, the TQ factorization is updated one row at a time. After the computation of each new row of T, a decision can be made (based on the size of the new diagonal) as to whether the constraint "should" be added. If the condition estimator is too large, the constraint is rejected. A rather conservative tolerance is used to reject constraints in this phase — for example, τ is not allowed to exceed $\epsilon^{-\frac{1}{2}}$, where ϵ is machine precision.

This strategy is enormously beneficial in situations where exact constraint dependencies have unknowingly (or perhaps deliberately) been introduced by the problem formulator. If the constraints are exactly dependent, the working set will include only a linearly independent subset. (An interesting side effect is that dependent *equality* constraints will never be included in the working set.) If the constraints are "nearly" dependent, the phase-1 procedure "hedges its bets" until the iterations begin, based on the hope that the offending constraints need never be added to the working set.

4.2. Adding constraints to the working set. Once iterations within the QP method have begun, the algorithm theoretically dictates precisely which constraint is to be added to the working set — namely, the "nearest" constraint reached by a step less than unity along the search direction. With exact arithmetic, there is no choice as to which constraint to add unless several constraints happen to intersect the search direction at exactly the same point (a highly unlikely occurrence). However, some flexibility can be introduced into the choice of constraint to be added *if we are prepared to tolerate "small" violations of constraints not in the working set.* We assume that the problem formulator specifies a vector δ of feasibility tolerances, one for each constraint. The i-th constraint is considered satisfied if the magnitude of the violation is less than δ_i. Even in the best possible circumstances, rounding errors imply that the violation will be of order machine precision. Therefore, δ_i is usually much larger than machine precision (say, of order $\epsilon^{\frac{1}{2}}$).

Consider a typical iteration of the QP method. To determine the constraint to be added, we define a set of "candidate" constraints. Let $\hat{\alpha}$ denote the maximum step such that all constraints not in the working set remain within their feasibility tolerances at $\hat{\alpha}$. The candidate constraints are those such that the *exact* step to the constraint is less than or equal to $\hat{\alpha}$. The constraint gradient a_i actually added to the working set corresponds to the candidate constraint whose normalized inner product with d is maximal, i.e., for which

$$\frac{|a_i^T d|}{\|a_i\|\|d\|}$$

is largest. Harris (1973) suggested this idea for improving numerical stability within the simplex method for linear programming.

The motivation for the procedure just described is the following. If a_i were linearly dependent on the working set, then $a_i^T d$ would be exactly zero. Therefore, if $|a_i^T d|$ is "small", a_i can be viewed as "nearly" linearly dependent on the working set. A "small" (normalized) value of $a_i^T d$ implies that a "large" step along d will tend to cause only a slight change in the residual. This constraint-addition strategy tends to generate an *indirect* control on the condition of the working set, and has led to significant improvements in performance on problems containing many nearly-dependent constraints.

It is interesting to note that any sensible test for controlling ill-conditioning (including those described above) will cope well with exact singularities, but will always be "wrong" in some situations of near-dependency. The example given by Fletcher (1981) of failure for SQP methods does not fail with the above strategies because the offending dependent constraint is never added to the working set. However, for "severely" ill-conditioned problems, any procedure will sometimes "fail" because of the inherent impossibility of problem-independent numerical rank estimation. (For a detailed discussion, see the classic paper by Peters and Wilkinson, 1970.)

4.3. Condition of the Hessian and projected Hessian. The approximation to the Hessian of the Lagrangian function is critical in obtaining favorable convergence of SQP methods (see, e.g., Powell, 1983). Within the QP subproblems, the important matrix is the projected Hessian matrix Z^THZ. In this section we consider algorithmic control of the condition of these matrices.

Information about the curvature of the Lagrangian function is represented in the $n \times n$ matrix H, a quasi-Newton approximation to the Hessian of the Lagrangian function. Much research has been devoted recently to techniques for defining H in SQP methods. In unconstrained optimization, the BFGS update has consistently been the most successful in practice (see, e.g., Dennis and Moré, 1977). One of the key features of the BFGS update is that it retains positive-definiteness of the approximate Hessian under suitable conditions on the step length α_k. On unconstrained problems, the line search can always be performed so that the updated matrix is guaranteed (at least in theory) to be positive definite. In practice, it is common to represent the Hessian approximation in terms of its Cholesky factors, which ensure numerical positive-definiteness and also permit explicit control of the condition estimate (the square of the ratio of the largest and smallest diagonals of the Cholesky factor).

For constrained problems, the situation is much more complicated. The Hessian of the Lagrangian function need not be positive definite anywhere, even at the solution. However, indefiniteness in H can lead to dire numerical and theoretical consequences, such as poorly posed QP subproblems and an inability to prove convergence. Therefore, most SQP methods maintain a positive-definite matrix H by some *modification* of the BFGS update. The BFGS update is

$$\bar{H} = H - \frac{1}{s^T H s} H s s^T H + \frac{1}{y^T s} y y^T, \tag{5}$$

where s is the change in x, and y is the change in gradient of the function whose Hessian is being approximated. If H is positive definite, a necessary and sufficient condition for $\bar{H}$ to be positive definite is that $y^T s > 0$. Since $y^T s$ may be negative for any choice of step length in an SQP method, y in (5) is taken as some other vector $\bar{y}$ such that $\bar{y}^T s > 0$ (see, e.g., Powell, 1977a). In practice, we have observed that y^T is nearly always positive; however, it is often *small*. Consequently, some modification of the approximate Hessian is frequently necessary to prevent H from becoming nearly singular.

The best choice of update for H is still the subject of active research. In our implementation, the Cholesky factors of H are updated after every major iteration, as in unconstrained optimization. Positive-definiteness is maintained by adding a perturbation in the range space of the active set. (For details, see Gill *et al.*, 1985a.)

Within the QP method, the projected Hessian matrix $Z^T H Z$ is also represented and updated in terms of its Cholesky factors. Even though the full Hessian H is ill-conditioned, the projected Hessian may remain well-conditioned throughout the QP. This has led many researchers to devise SQP methods in which only an approximation to the projected Hessian is retained (see, e.g., Murray and Wright, 1978; Coleman and Conn, 1982; Nocedal and Overton, 1983).

5. Numerical examples

In this section, we give two examples in order to illustrate some of the issues mentioned in previous sections. The printed output requires some preliminary explanation. Each major iteration generates a single line of output. The major iteration number is given in the first column (marked "ITN"). The next column "ITQP" gives the number of minor iterations needed to solve the QP subproblem. The "STEP" column gives the step α_k taken along the computed search direction. "NUMF" is the total number of evaluations of the problem functions. The merit function value is given by "MERIT". Columns "BND" and "LC" give the numbers of simple-bound constraints and general linear constraints in the working set; columns "NC", "NORM C" and "RHO" give the number of nonlinear constraints in the working set, the two-norm of the residuals of constraints in the working set and the penalty parameter used in the merit function. "NZ" is the dimension of the null space of the current matrix of constraints in the working set. The next five entries give information about the derivatives of the problem at the current point. "NORM GF" is the two-norm of the free components of the objective gradient g_k, and "NORM GZ" is the two-norm of $Z_k^T g_k$. "COND H", "COND HZ" and "COND T" are estimates of the condition numbers of the Hessian, projected Hessian and matrix of constraints in the working set. "CONV" is a set of three logical variables C_1, C_2 and C_3, that indicate properties of the current estimate of the solution, with the following meanings. C_1 is true if the projected-gradient norm is small; C_2 is true if constraints are satisfied to within the user-specified tolerance; and C_3 is true if the last change in x was small.

A value of "1" for ITQP in the final iterations indicates that the correct active set has been identified. On difficult problems (such as the second example), the predicted active set may not "settle down" until near the end of the run. Indications of the expected superlinear convergence rate are unit steplengths in the "STEP" column and the sequence of diminishing "NORM GZ" entries.

The final solution printout is divided into three sections, giving information about the final status of the variables, general linear constraints and nonlinear constraints. Within each section, "STATE" gives the status of the associated constraint in the predicted active set ("FR" if not included, "EQ" if a fixed value, "LL" if at its lower bound, and "UL" if at its upper bound). "VALUE" is the value of the constraint at the final iteration. "LOWER BOUND" and "UPPER BOUND" give the lower and upper bounds specified for the constraint ("NONE" indicates that the bound is infinite). "LAGR MULTIPLIER" is the value of the Lagrange multiplier. This will be zero if STATE is FR. The multiplier is non-negative if STATE is LL, and non-positive if STATE is UL. "RESIDUAL" gives the difference between the entry in the VALUE column and the nearer bound.

All computation was performed in double precision on an IBM 3081, which corresponds to $\epsilon \approx 10^{-16}$. The feasibility tolerances were set to 10^{-8} for the linear constraints and 10^{-6} for the nonlinear constraints.

Figure 1 gives the results obtained on the "Powell triangles" problem (see Powell, 1977b). The problem contains seven variables, four non-infinite bounds and five nonlinear constraints. The Hessian of the Lagrangian function is not positive definite at the solution.

As is typical of well-behaved problems, the Hessian approximation and working set remain relatively well-conditioned. Similarly, the penalty parameter remains small. As the iterates converge, only one minor iteration is performed per major iteration, and entries in the "NORM GZ"

ITN	ITQP	STEP	NUMF	MERIT	BND	LC	NC	NZ	NORM GF	NORM GZ	COND H	COND HZ	COND T	NORM C	RHO	CONV
0	6	0.0E+00	1	6.0000E+00	2	0	4	1	3.6E+00	6.07E-01	1.E+00	1.E+00	2.E+01	3.87E+01	0.0E+00	FFT
1	2	1.0E+00	2	2.4092E+01	1	0	4	2	6.7E+00	1.16E+00	2.E+00	1.E+00	2.E+01	8.91E+00	5.6E+00	FFF
2	1	1.0E+00	3	2.7313E+01	1	0	4	2	8.2E+00	1.31E+00	2.E+02	2.E+00	1.E+01	2.50E+00	3.9E+00	FFF
3	1	1.0E+00	4	2.5333E+01	1	0	4	2	8.1E+00	1.38E+00	5.E+03	2.E+00	9.E+00	2.41E-01	3.9E+00	FFF
4	1	1.0E+00	5	2.5062E+01	1	0	4	2	7.4E+00	8.68E-01	1.E+02	4.E+00	7.E+00	4.54E-01	3.9E+00	FFF
5	1	1.0E+00	6	2.4436E+01	1	0	4	2	7.7E+00	8.27E-01	2.E+02	3.E+00	8.E+00	8.39E-02	4.1E+00	FFF
6	1	1.0E+00	7	2.4383E+01	1	0	4	2	7.7E+00	8.53E-01	2.E+03	2.E+00	9.E+00	1.43E-03	1.6E+01	FFF
7	1	1.0E+00	8	2.3922E+01	1	0	4	2	7.3E+00	6.93E-01	2.E+02	1.E+00	7.E+00	1.92E-02	1.6E+01	FFF
8	1	1.0E+00	9	2.3364E+01	1	0	4	2	6.8E+00	2.30E-01	9.E+01	4.E+00	8.E+00	9.98E-02	1.6E+01	FFF
9	1	6.1E-01	11	2.3315E+01	1	0	4	2	6.8E+00	6.54E-02	1.E+03	3.E+00	9.E+00	6.90E-02	1.7E+01	FFF
10	1	1.0E+00	12	2.3315E+01	1	0	4	2	6.8E+00	3.63E-02	2.E+03	2.E+00	9.E+00	7.49E-03	2.1E+01	FFF
11	1	1.0E+00	13	2.3314E+01	1	0	4	2	6.8E+00	2.18E-02	2.E+02	1.E+00	6.E+00	2.93E-05	2.1E+01	FFF
12	1	1.0E+00	14	2.3314E+01	1	0	4	2	6.8E+00	3.11E-03	9.E+02	1.E+00	7.E+00	9.22E-05	2.1E+01	FFF
13	1	1.0E+00	15	2.3314E+01	1	0	4	2	6.8E+00	1.23E-03	2.E+03	1.E+00	9.E+00	1.28E-06	2.1E+01	FTF
14	1	1.0E+00	16	2.3314E+01	1	0	4	2	6.8E+00	4.69E-05	2.E+03	1.E+00	9.E+00	1.01E-06	2.1E+01	TTF
15	1	1.0E+00	17	2.3314E+01	1	0	4	2	6.8E+00	2.88E-06	2.E+03	1.E+00	9.E+00	1.08E-09	2.1E+01	TTF
16	1	1.0E+00	18	2.3314E+01	1	0	4	2	6.8E+00	2.97E-08	8.E+02	1.E+00	7.E+00	1.84E-12	2.1E+01	TTT

EXIT NP PHASE. INFORM = 0 MAJITS = 16 NFEVAL = 18 NCEVAL = 18

VARIABLE	STATE	VALUE	LOWER BOUND	UPPER BOUND	LAGR MULTIPLIER	RESIDUAL
VARBL 1	FR	4.828427	0.0000000E+00	NONE	0.0000000E+00	4.828
VARBL 2	FR	-0.3645023E-07	NONE	NONE	0.0000000E+00	0.1000E+11
VARBL 3	FR	4.828427	0.0000000E+00	NONE	0.0000000E+00	4.828
VARBL 4	FR	1.000000	NONE	NONE	0.0000000E+00	0.1000E+11
VARBL 5	FR	2.414214	1.000000	NONE	0.0000000E+00	1.414
VARBL 6	FR	2.414214	NONE	NONE	0.0000000E+00	0.1000E+11
VARBL 7	LL	1.000000	1.000000	NONE	9.656854	0.0000E+00

NONLNR CONSTR	STATE	VALUE	LOWER BOUND	UPPER BOUND	LAGR MULTIPLIER	RESIDUAL
NLCON 1	LL	0.1458611E-11	0.0000000E+00	NONE	1.707107	0.1459E-11
NLCON 2	LL	-0.1044248E-11	0.0000000E+00	NONE	9.656854	-0.1044E-11
NLCON 3	LL	0.2418066E-12	0.0000000E+00	NONE	6.828427	0.2418E-12
NLCON 4	FR	1.414214	0.0000000E+00	NONE	0.0000000E+00	1.414
NLCON 5	LL	-0.3489570E-12	0.0000000E+00	NONE	6.828427	-0.3490E-12

EXIT NPSOL - OPTIMAL SOLUTION FOUND.

FINAL NONLINEAR OBJECTIVE VALUE = 23.31371

Figure 1. Output from the well-behaved problem "Powell triangles".

and "NORM C" columns exhibit superlinear convergence. Note that the accuracy of the nonlinear constraints is considerably better than the projected-gradient norm for several iterations before termination. Another feature of interest is that the the constraint values and Lagrange multipliers at the solution are "well balanced". For example, all the multipliers are of approximately the same order of magnitude. This behavior is typical of a well-scaled problem.

The second example is the problem "Dembo 7", which is a geometric programming formulation developed by Dembo (1976) of a five-stage membrane separation process. The problem has sixteen variables, eight linear constraints, and eleven nonlinear constraints. All sixteen variables have simple upper and lower bound constraints. The problem is notoriously difficult because of bad scaling and nearly dependent constraints.

The results for Dembo 7 show a number of features typical of badly behaved problems. First, note that, in contrast to the first example, the number of minor iterations does not decrease

ITN	ITQP	STEP	NUMF	MERIT	BND	LC	NC	NZ	NORM GF	NORM GZ	COND H	COND HZ	COND T	NORM C	RHO	CONV
0	27	0.0E+00	1	2.8459E+02	1	3	9	3	1.1E+03	5.17E-01	1.E+00	1.E+00	7.E+03	2.19E-01	0.0E+00	FFT
1	13	9.6E-01	4	-2.1324E+02	3	2	5	6	8.1E+02	1.76E+00	2.E+03	4.E+01	3.E+02	1.50E+00	1.4E+06	FFF
2	12	8.1E-03	6	3.7305E+02	2	2	7	5	8.1E+02	1.87E+00	4.E+07	1.E+00	1.E+05	1.49E+00	1.6E+03	FFF
3	12	5.7E-02	8	4.4182E+02	2	1	7	6	8.1E+02	1.12E+02	4.E+04	2.E+02	5.E+03	1.41E+00	8.0E+02	FFF
4	7	1.0E+00	9	4.0201E+02	2	3	5	6	8.2E+02	2.25E+00	2.E+01	1.E+00	4.E+02	1.86E-01	8.0E+02	FFF
5	9	1.0E+00	10	3.9278E+02	2	1	7	6	8.1E+02	1.62E+00	2.E+03	2.E+01	8.E+01	4.18E-01	1.3E+03	FFF
6	17	1.0E+00	11	3.0388E+02	2	3	5	6	1.7E+02	7.24E-01	4.E+08	2.E+08	6.E+03	3.03E-01	8.8E+01	FFF
7	29	1.0E+00	12	1.7889E+02	2	3	6	5	6.3E+02	3.03E+01	2.E+10	8.E+03	2.E+03	8.89E-01	3.7E+01	FFF
8	13	1.0E+00	13	1.8780E+02	1	3	7	5	6.3E+02	3.10E+01	2.E+09	7.E+05	3.E+03	2.13E-02	1.1E+03	FFF
9	5	1.0E+00	14	1.8768E+02	2	4	7	3	6.3E+02	2.76E+01	2.E+09	1.E+03	9.E+04	1.43E-01	1.6E+02	FFF
10	1	1.0E+00	15	1.8535E+02	2	4	7	3	6.3E+02	2.47E+01	5.E+09	3.E+03	3.E+04	3.67E-04	1.6E+02	FFF
11	5	1.0E+00	16	1.8509E+02	2	3	6	5	6.3E+02	8.17E+01	6.E+11	1.E+06	1.E+05	6.62E-02	1.6E+02	FFF
12	4	1.0E+00	17	1.8001E+02	3	3	6	4	6.2E+02	2.66E+01	1.E+11	1.E+05	5.E+05	9.14E-03	1.6E+02	FFF
13	2	1.0E+00	18	1.7794E+02	4	3	5	4	6.2E+02	2.67E+01	4.E+10	4.E+02	7.E+03	6.41E-03	1.6E+02	FFF
14	5	1.0E+00	19	1.7761E+02	3	4	5	4	6.2E+02	2.79E+01	1.E+12	3.E+04	2.E+03	7.13E-02	1.6E+02	FFF
15	9	1.0E+00	20	1.7469E+02	3	3	6	4	6.2E+02	3.58E+00	1.E+12	5.E+05	2.E+03	7.62E-03	1.6E+02	FFF
16	7	2.7E-01	22	1.7507E+02	3	3	6	4	6.2E+02	3.93E+00	6.E+11	3.E+05	2.E+03	5.61E-03	1.7E+04	FFF
17	7	1.0E+00	23	1.7509E+02	3	3	6	4	6.2E+02	4.51E+00	3.E+11	1.E+05	2.E+03	3.38E-04	2.1E+04	FFF
18	2	1.0E+00	24	1.7508E+02	4	3	6	3	6.2E+02	4.53E+00	3.E+11	2.E+05	3.E+03	5.69E-01	1.3E+03	FFF
19	1	1.0E+00	25	1.7493E+02	4	3	6	3	6.2E+02	4.36E+00	5.E+10	5.E+03	3.E+03	3.92E-05	5.0E+02	FFF
20	2	1.0E+00	26	1.7493E+02	4	4	6	2	6.2E+02	4.29E+00	5.E+11	8.E+04	4.E+03	1.27E-02	5.0E+02	FFF
21	4	1.0E+00	27	1.7485E+02	4	3	6	3	6.2E+02	3.58E+00	9.E+11	9.E+06	4.E+03	4.03E-04	5.0E+02	FFF
22	2	1.0E+00	28	1.7483E+02	4	4	6	2	6.2E+02	3.60E+00	6.E+11	7.E+04	4.E+03	7.88E-04	5.0E+02	FFF
23	1	1.0E+00	29	1.7483E+02	4	4	6	2	6.2E+02	3.53E+00	9.E+11	2.E+05	4.E+03	1.75E-06	5.0E+02	FTF
24	2	1.0E+00	30	1.7483E+02	5	4	6	1	6.2E+02	1.84E-03	8.E+13	1.E+00	9.E+03	3.03E-03	5.0E+02	TFT
25	4	4.8E-01	32	1.7482E+02	4	3	7	2	6.2E+02	9.67E-03	3.E+13	4.E+01	6.E+03	1.49E-02	5.0E+02	FFF
26	1	1.0E+00	33	1.7480E+02	4	3	7	2	6.2E+02	1.44E-02	2.E+12	3.E+01	1.E+04	4.35E-05	5.0E+02	FFF
27	1	1.0E+00	34	1.7480E+02	4	3	7	2	6.2E+02	1.19E-02	2.E+12	2.E+02	1.E+04	2.57E-06	5.0E+02	FTF
28	1	1.0E+00	35	1.7479E+02	4	3	7	2	6.2E+02	2.96E-03	9.E+11	8.E+01	1.E+04	1.06E-04	5.0E+02	TFF
29	1	1.0E+00	36	1.7479E+02	4	3	7	2	6.2E+02	2.08E-03	1.E+12	8.E+01	1.E+04	1.20E-05	5.0E+02	TTF
30	1	1.0E+00	37	1.7479E+02	4	3	7	2	6.2E+02	1.71E-03	2.E+11	1.E+02	1.E+04	6.85E-08	5.0E+02	TTF
31	1	1.0E+00	38	1.7479E+02	4	3	7	2	6.2E+02	6.66E-04	2.E+11	2.E+02	1.E+04	5.48E-07	5.0E+02	TTF
32	1	1.0E+00	39	1.7479E+02	4	3	7	2	6.2E+02	2.89E-04	2.E+11	2.E+02	1.E+04	9.15E-08	5.0E+02	TTF
33	1	1.0E+00	40	1.7479E+02	4	3	7	2	6.2E+02	2.89E-04	2.E+12	1.E+01	2.E+04	1.97E-12	5.0E+02	TTT

EXIT NP PHASE. INFORM = 0 MAJITS = 33 NFEVAL = 40 NCEVAL = 40

VARIABLE	STATE	VALUE	LOWER BOUND	UPPER BOUND	LAGR MULTIPLIER	RESIDUAL
VARBL 1	FR	0.8036427	0.1000000	0.9000000	0.0000000E+00	0.9636E-01
VARBL 2	FR	0.8158941	0.1000000	0.9000000	0.0000000E+00	0.8411E-01
VARBL 3	FR	0.9000000	0.1000000	0.9000000	0.0000000E+00	0.1388E-15
VARBL 4	UL	0.9000000	0.1000000	0.9000000	-813.4083	0.0000E+00
VARBL 5	LL	0.9000000	0.9000000	1.000000	392.3255	0.0000E+00
VARBL 6	FR	0.9992972E-01	0.1000000E-03	0.1000000	0.0000000E+00	0.7028E-04
VARBL 7	FR	0.1069025	0.1000000	0.9000000	0.0000000E+00	0.6903E-02
VARBL 8	FR	0.1908367	0.1000000	0.9000000	0.0000000E+00	0.9084E-01
VARBL 9	FR	0.1908367	0.1000000	0.9000000	0.0000000E+00	0.9084E-01
VARBL 10	FR	0.1908367	0.1000000	0.9000000	0.0000000E+00	0.9084E-01
VARBL 11	FR	499.9135	1.000000	1000.000	0.0000000E+00	498.9
VARBL 12	FR	4.953426	0.1000000E-05	500.0000	0.0000000E+00	4.953
VARBL 13	FR	72.66272	1.000000	500.0000	0.0000000E+00	71.66
VARBL 14	FR	500.0000	500.0000	1000.000	0.0000000E+00	0.2842E-12
VARBL 15	LL	500.0000	500.0000	1000.000	0.9018375	0.0000E+00
VARBL 16	LL	0.1000000E-05	0.1000000E-05	500.0000	0.1547569	0.1000E-17

LINEAR CONSTR	STATE	VALUE	LOWER BOUND	UPPER BOUND	LAGR MULTIPLIER	RESIDUAL
LNCON 1	FR	-494.9601	NONE	0.0000000E+00	0.0000000E+00	495.0
LNCON 2	FR	0.0000000E+00	NONE	0.0000000E+00	0.0000000E+00	0.0000E+00
LNCON 3	UL	-0.1387779E-15	NONE	0.0000000E+00	-724.6329	0.1388E-15
LNCON 4	FR	-0.8410590E-01	NONE	0.0000000E+00	0.0000000E+00	0.8411E-01
LNCON 5	FR	-0.1225144E-01	NONE	0.0000000E+00	0.0000000E+00	0.1225E-01
LNCON 6	UL	0.0000000E+00	NONE	0.0000000E+00	-290.3208	0.0000E+00
LNCON 7	UL	-0.1942890E-15	NONE	0.0000000E+00	-290.3227	0.1943E-15
LNCON 8	FR	0.9899202	NONE	1.000000	0.0000000E+00	0.1008E-01

NONLNR CONSTR	STATE	VALUE	LOWER BOUND	UPPER BOUND	LAGR MULTIPLIER	RESIDUAL
NLCON 1	LL	-0.8326673E-16	0.0000000E+00	NONE	6.901206	-0.8327E-16
NLCON 2	LL	-0.3939904E-12	0.0000000E+00	NONE	102.6914	-0.3940E-12
NLCON 3	FR	0.1994932E-16	0.0000000E+00	NONE	0.0000000E+00	0.1995E-16
NLCON 4	FR	0.1127570E-16	0.0000000E+00	NONE	0.0000000E+00	0.1128E-16
NLCON 5	LL	0.1127570E-16	0.0000000E+00	NONE	368.3360	0.1128E-16
NLCON 6	LL	-0.1924860E-11	0.0000000E+00	NONE	1.733360	-0.1925E-11
NLCON 7	LL	-0.5504017E-13	0.0000000E+00	NONE	40.15522	-0.5504E-13
NLCON 8	LL	-0.2738088E-13	0.0000000E+00	NONE	417.6574	-0.2738E-13
NLCON 9	LL	-0.1249001E-15	0.0000000E+00	NONE	474.0819	-0.1249E-15
NLCON 10	FR	0.1110223E-15	0.0000000E+00	NONE	0.0000000E+00	0.1110E-15
NLCON 11	FR	0.1665335E-15	0.0000000E+00	NONE	0.0000000E+00	0.1665E-15

EXIT NPSOL - OPTIMAL SOLUTION FOUND.

Figure 2. Output from the solution of Dembo 7.

quickly. Moreover, the presence of near-zero Lagrange multipliers sometimes causes more than one QP iteration to be required even relatively close to the solution. A common symptom of a difficult problem is a large value of the condition estimator of the full approximate Hessian, which contrasts with the relatively modest condition of the *projected* Hessian.

The significant variation in size of the active constraints indicates poor scaling of the constraints, which implies that the algorithm may have difficulty in identifying the active set. In fact, the third bound constraint, the third linear constraint and the eleventh nonlinear constraint all have very small residuals, but are not in the active set of the final QP subproblem. In constrast to the first example, in which the accuracy of the constraints was much better than the convergence tolerance, some of the nonlinear constraints barely satisfy the required feasibility tolerance.

Finally, we wish to emphasize that, despite severe ill-conditioning in the Hessian of the Lagrangian and serious dependencies among the constraints, Dembo 7 is solved in a relatively routine manner. As discussed in Sections 4.1 and 4.2, dependent constraints are successfully omitted from the working set so that its condition estimator never becomes too large.

Acknowledgement

This research was supported by the U.S. Army Research Office Contract DAAG29-84-K-0156, the U.S. Department of Energy Contract DE-AM03-76SF00326, PA No. DE-AT03-76ER72018; National Science Foundation Grants MCS-7926009 and ECS-8312142; and the Office of Naval Research Contract N00014-75-C-0267.

References

Coleman, T. F. and Conn, A. R. (1982). Nonlinear programming via an exact penalty function: asymptotic analysis, *Mathematical Programming* **24**, pp. 123–136.

Coleman, T. F. and Conn, A. R. (1984). On the local convergence of a quasi-Newton method for the nonlinear programming problem, *SIAM Journal on Numerical Analysis* **21**, pp. 755–769.

Coleman, T. F. and Sorensen, D. C. (1984). A note on the computation of an orthogonal basis for the null space of a matrix, *Mathematical Programming* **29**, pp. 234–242.

Dembo, R. S. (1976). A set of geometric test problems and their solutions, *Mathematical Programming* **10**, pp. 192–213.

Dennis, J. E., Jr. and Moré, J. J. (1977). Quasi-Newton methods, motivation and theory, *SIAM Review* **19**, pp. 46–89.

Fletcher, R. (1981). "Numerical experiments with an exact ℓ_1 penalty function method", in *Nonlinear Programming 4* (O. L. Mangasarian, R. R. Meyer and S. M. Robinson, eds.), Academic Press, London and New York, pp. 99–129.

Gill, P. E., Murray, W., Saunders, M. A. and Wright, M. H. (1983). On the representation of a basis for the null space, Report SOL 83–19, Department of Operations Research, Stanford University, Stanford, California.

Gill, P. E., Murray, W., Saunders, M. A. and Wright, M. H. (1984a). User's guide to SOL/NPSOL (Version 2.1), Report SOL 84–7, Department of Operations Research, Stanford University, Stanford, California.

Gill, P. E., Murray, W., Saunders, M. A. and Wright, M. H. (1984b). Procedures for optimization problems with a mixture of bounds and general linear constraints, *ACM Transactions on Mathematical Software* **10**, pp. 282–298.

Gill, P. E., Murray, W., Saunders, M. A. and Wright, M. H. (1984c). User's guide to SOL/QPSOL (Version 3.2), Report SOL 84–6, Department of Operations Research, Stanford University, Stanford, California.

Gill, P. E., Murray, W., Saunders, M. A. and Wright, M. H. (1985a). Model building and practical aspects of nonlinear programming, Report SOL 85–3, Department of Operations Research, Stanford University, Stanford, California.

Gill, P. E., Murray, W., Saunders, M. A. and Wright, M. H. (1985b). The design and implementation of a quadratic programming algorithm, to appear.

Gill, P. E., Murray, W., Saunders, M. A., Stewart, G. W. and Wright, M. H. (1985c). Properties of a representation of a basis for the null space, Report SOL 85–1, Department of Operations Research, Stanford University, Stanford, California.

Gill, P. E., Murray, W. and Wright, M. H. (1981). *Practical Optimization*, Academic Press, London and New York.

Harris, P. M. J. (1973). Pivot selection methods of the Devex LP code, *Mathematical Programming* **5**, pp. 1–28. [Reprinted in *Mathematical Programming Study* **4** (1975), pp. 30–57.]

Murray, W. and Wright, M. H. (1978). Methods for nonlinearly constrained optimization based on the trajectories of penalty and barrier functions, Report SOL 78–23, Department of Operations Research, Stanford University, Stanford, California.

Nocedal, J. and Overton, M. L. (1983). Projected Hessian updating algorithms for nonlinearly constrained optimization, Report 95, Department of Computer Science, Courant Institute of Mathematical Sciences, New York University, New York, New York.

Ortega, J. M. and Rheinboldt, W. C. (1970). *Iterative Solution of Nonlinear Equations in Several Variables*, Academic Press, London and New York.

Peters, G. and Wilkinson, J. H. (1970). The least-squares problem and pseudo-inverses, *Computer Journal* **13**, pp. 309–316.

Powell, M. J. D. (1974). "Introduction to constrained optimization", in *Numerical Methods for Constrained Optimization* (P. E. Gill and W. Murray, eds.), pp. 1–28, Academic Press, London and New York.

Powell, M. J. D. (1977a). A fast algorithm for nonlinearly constrained optimization calculations, Report DAMTP 77/NA 2, University of Cambridge, England.

Powell, M. J. D. (1977b). Variable metric methods for constrained optimization, Report DAMTP 77/NA 5, University of Cambridge, England.

Powell, M. J. D. (1983). "Variable metric methods for constrained optimization", in *Mathematical Programming: The State of the Art*, (A. Bachem, M. Grötschel and B. Korte, eds.), pp. 288–311, Springer-Verlag, Berlin, Heidelberg, New York and Tokyo.

Robinson, S. M. (1974). Perturbed Kuhn-Tucker points and rates of convergence for a class of nonlinear programming algorithms, *Math. Prog.* **7**, pp. 1–16.

Schittkowski, K. (1981). The nonlinear programming method of Wilson, Han and Powell with an augmented Lagrangian type line search function, *Numerische Mathematik* **38**, pp. 83–114.

Schittkowski, K. (1982). On the convergence of a sequential quadratic programming method with an augmented Lagrangian line search function, Report SOL 82-4, Department of Operations Research, Stanford University, Stanford, California.

Tone, K. (1983). Revisions of constraint approximations in the successive QP method for nonlinear programming, *Mathematical Programming* **26**, pp. 144–152.

Wilson, R. B. (1963). A Simplicial Algorithm for Concave Programming, Ph. D. Thesis, Harvard University.

REMARKS ON A CONTINUOUS FINITE ELEMENT SCHEME FOR HYPERBOLIC EQUATIONS

Richard S. Falk
Department of Mathematics
Rutgers University
New Brunswick, NJ 08903

Gerard R. Richter
Department of Computer Science
Rutgers University
New Brunswick, NJ 08903

1. Introduction

Hyperbolic partial differential equations are most commonly discretized by finite difference methods. When a finite element method is applied to a hyperbolic problem, it is often applied only in the spatial variables, leaving a system of time dependent ordinary differential equations to be solved numerically (usually by finite difference methods). Examples of this approach can be found in [1] and [3].

An alternative approach is to obtain a full finite element discretization of the problem, without distinguishing one of the independent variables for special treatment. It is possible to do this in such a way that the resulting discretization is essentially explicit and incorporates the initial/boundary data in a manner that is natural to the problem. Several such schemes, applicable over a triangulation of a domain in R^2, were introduced by Reed and Hill [7]. One produces a discontinuous piecewise polynomial approximation and was analyzed by Lesaint and Raviart [6], with improved estimates later obtained by Johnson and Pitkaranta [5]. Two others use different test spaces to produce continuous piecewise polynomial approximations. One of these continuous schemes is the focal point of this paper. Other work in this direction has been done by Winther [8], who obtained optimal order error estimates for a continuous finite element method applicable over a rectangular mesh.

We shall confine our attention here to the model problem:

$$\alpha \cdot \nabla u + \beta u = f \quad \text{in } \Omega \tag{1}$$

$$u = g \quad \text{on the inflow boundary } \Gamma_{in}(\Omega),$$

where α is a constant unit vector, β is a constant, and Ω is a bounded polygonal domain in R^2.

The analysis presented here amounts to a condensed version of [4], to which the interested reader is referred for details of proofs and a treatment of the more general case of variable coefficients.

2. Description of the Method

To describe the method of interest in this paper, we let Δ_h be a quasiuniform triangulation of Ω, indexed by the maximum triangle side h, and constructed so that no triangle has a side parallel to the characteristic direction. For any subdomain Ω_S of Ω, we denote by $\Gamma_{in}(\Omega_S)$ the inflow portion of the boundary of Ω_S, i.e., $\{ x \in \Gamma(\Omega_S) \mid \alpha \cdot n < 0 \}$, where n is the unit outward normal to Ω_S, and by $\Gamma_{out}(\Omega_S)$ the remaining (outflow) portion of $\Gamma(\Omega_S)$. With Δ_h as above, each triangle has one inflow side and two outflow sides (a *type I* triangle) or two inflow sides and one outflow side (a *type II* triangle). Furthermore, the triangles $\{T_i\}$ in Δ_h may be ordered so that the domain of dependence of T_k contains none of $T_{k+1}, T_{k+2}, \ldots$ (see [6]). This ordering allows one to develop an approximate solution in an explicit manner, first in T_1, then in T_2, etc. At the point when the solution is to be formed in a given triangle, it will be known along the inflow to that triangle.

We seek an approximate solution in the subspace

$$S_h^n = \{ v_h \in C^0(\Omega) \text{ such that } v_h|_T \in \mathbf{P}_n(T) \},$$

where $n \geq 2$ and $\mathbf{P}_n(T)$ denotes the space of polynomials of degree $\leq n$ over the triangle T. Letting g_I be a suitable interpolant of g in $S_h^n|_{\Gamma_{in}(\Omega)}$, and denoting the L^2 inner product over T by $(\ ,\ )_T$, the finite element method we consider can be described as follows:

Problem $\mathbf{P_h}$: Find $u_h \in S_h^n$ such that $u_h = g_I$ on $\Gamma_{in}(\Omega)$, and for triangles of type I

$$(\alpha \cdot \nabla u_h + \beta u_h, v_h)_T = (f, v_h)_T \quad \text{for all } v_h \in \mathbf{P}_{n-1}(T), \tag{2}$$

while for triangles of type II

$$(\alpha \cdot \nabla u_h + \beta u_h, v_h)_T = (f, v_h)_T \quad \text{for all } v_h \in \mathbf{P}_{n-2}(T). \tag{3}$$

Note that the approximate solution u_h has a total of $\sigma_n \equiv \sum_{j=1}^{n+1} j$ degrees of freedom in each triangle. In a one-inflow-side triangle, there are $n+1$ degrees of freedom in u_h along the inflow, leaving a total of σ_{n-1} to be determined from equation (2). In a two-inflow-side triangle, there are $2n+1$ degrees of freedom in u_h along the inflow, leaving σ_{n-2} to be determined from equation (3). Thus in both equations (2) and (3), the number of equations equals the number of unknowns.

For example, when $n=2$, we are approximating u by a C^0 piecewise quadratic. Hence u_h is determined by its values at the vertices of the triangulation and at the midpoints of the triangle sides. In a one-inflow-side triangle, the three degrees of freedom at the midpoint and two vertices of the inflow side are already known. The three equations produced from (2) by taking $v_h = 1, x, y$ then determine the approximate solution at the third vertex and two remaining midpoints. In a two-inflow-side triangle, the only degree of freedom remaining is the value of u_h at the midpoint of the outflow side. This is determined from equation (3) by taking $v_h = 1$.

For computational and theoretical purposes, it is advantageous to think of the triangles in Δ_h as partioned into *layers* S_i. We define these as follows:

$$S_1 = \{ T \in \Delta_h \,|\, \Gamma_{in}(T) \subseteq \Gamma_{in}(\Omega) \}$$

$$S_{i+1} = \{ T \in \Delta_h \,|\, \Gamma_{in}(T) \subseteq \Gamma_{in}(\Omega - \cup_{k\leq i} S_k) \}, \quad i = 1, 2, \ldots$$

With this partition of Δ_h, the approximate solution may be obtained in an explicit manner, first in S_1, then in S_2, etc. Within each layer, the approximate solution can be obtained in parallel since the solution in any of the triangles within a layer does not depend on the solution in other triangles in that layer.

The following additional notation will be used in the subsequent sections. For D a domain in R^2, $\|f\|_D \equiv (\int_D f^2 \, dx\, dy)^{1/2}$, and for Γ a line segment, $|f|_\Gamma \equiv (\int_\Gamma f^2 \, d\tau)^{1/2}$. We denote by $\|f\|_{n,D}$ the norm in the Sobolev space $H^n(D)$. When $D = \Omega$, we shall omit the subscript D. We shall also denote by $P_k f$ the L^2 projection over T into $\mathbf{P}_k(T)$ (the space of polynomials of degree $\leq k$ over T). For α a constant unit vector, we shall use the notation u_α to mean $\nabla u \cdot \alpha$. It will also be convenient to have the following notation relative to an arbitrary triangle T of Δ_h. For $i = 1, 2, 3$, we denote by Γ_i the sides of T numbered counterclockwise, by a_i the vertices of T opposite Γ_i, by n_i the unit outward normals to Γ_i, and by τ_i the unit tangents along Γ_i taken in a counterclockwise direction. *We shall always take Γ_3 to be the inflow side of of a type I triangle or the outflow side of a type II triangle.* Finally, we shall use the symbol C to denote a generic constant, independent of u and h.

3. Existence and Uniqueness

By a scaling argument, one can show that if Problem P_h with $\beta = 0$ has a unique solution, so does the general problem, provided h is sufficiently small. For this simplified problem, proving existence and uniqueness is equivalent to showing that the only solution to equations (2) or (3) with $\beta = f = 0$, and $u_h = 0$ on $\Gamma_{in}(T)$ is $u_h \equiv 0$. We can show this as follows:

For type I triangles, we take $v_h = (u_h)_\alpha \in \mathbf{P}_{n-1}(T)$, and infer that $(u_h)_\alpha = 0$. This, together with $u_h = 0$ on $\Gamma_{in}(T)$, imply that $u_h \equiv 0$ in T.

For type II triangles, we note that $u_h = 0$ on $\Gamma_{in}(T)$ implies that u_h can be written in the form $u_h = \xi\eta w_h$ where ξ and η are coordinates along the two inflow sides, with $\xi = \eta = 0$ at a_3, $\xi,\eta \geq 0$ in T, and $w_h \in \mathbf{P}_{n-2}(T)$. Taking $v_h = w_h$ in (3), we integrate by parts to obtain:

$$0 = ((u_h)_\alpha, w_h)_T = (\xi\eta(w_h)_\alpha, w_h)_T + ((\xi\eta)_\alpha w_h, w_h)_T$$

$$= \frac{1}{2}(\xi\eta, (w_h^2)_\alpha)_T + ((\xi\eta)_\alpha, w_h^2)_T$$

$$= \frac{1}{2}\int_{\Gamma_{out}(T)} w_h^2 \xi\eta\, \alpha\cdot n\, d\Gamma + \frac{1}{2}((\xi\eta)_\alpha, w_h^2)_T.$$

Now $(\xi\eta)_\alpha$ is positive in T and $\xi\eta\, \alpha\cdot n$ is nonnegative on $\Gamma_{out}(T)$. Hence $w_h \equiv 0$ in T and $u_h \equiv 0$ in T.

4. Basic Identities

The test function $v_h = -(u_h)_{\tau_1\tau_2}$ in (2) and (3) will play a key role in our stability analysis, where τ_1 and τ_2 are the tangents to the two outflow sides of a type I triangle or the two inflow sides of a type II triangle. Following are equivalent expressions for the two terms in the integral

$$(u_\alpha + \beta u, -u_{\tau_1\tau_2})_T\,.$$

Proofs may be found in [4].

Lemma 4.1: For any constant unit vector α and any twice differentiable function u:

$$\int_T u_\alpha(-u_{\tau_1\tau_2})\,dx\,dy = \frac{1}{2}\int_{\Gamma(T)} \frac{(\alpha\cdot n_1)(\alpha\cdot n_2)}{\alpha\cdot n} u_\tau^2\, d\tau - \frac{1}{2}\int_{\Gamma_3} \frac{(\tau_1\cdot n_3)(\tau_2\cdot n_3)}{\alpha\cdot n_3} u_\alpha^2\, d\tau$$

Lemma 4.2: Let $T \in \{\Delta_h\}$, and for $P \in \Gamma$ let $\theta(P)$ be the angle from $\alpha(P)$ to the local tangent vector τ, measured counterclockwise. Then

$$\int_T \beta u(-u_{\tau_1\tau_2})\,dx\,dy = \int_T \beta u_{\tau_1} u_{\tau_2}\,dx\,dy - \int_{\Gamma(T)} \beta(\alpha\cdot n_1)(\alpha\cdot n_2)\cot\theta\, u\, u_\tau\, d\tau$$

$$- \int_{\Gamma_3(T)} \beta \frac{(\tau_2\cdot n_3)(\tau_1\cdot n_3)}{\alpha\cdot n_3} u\, u_\alpha\, d\tau$$

5. Stability

We now indicate the steps in deriving the basic stability estimates for Problem $\mathbf{P_h}$. These will be used to obtain error estimates in the next section. The first step is to develop local stability results applicable over a single triangle. This is complicated somewhat by the fact that the two different types of triangles require different treatment. For each, we shall obtain a bound on the growth of $\frac{du_h}{d\tau}$ from the identities in the previous section. When these are combined suitably with bounds on the growth of u_h, the desired stability result is obtained.

The next two lemmas bound the growth in u_h over the two types of triangles.

Lemma 5.1: If T is a type I triangle and u_h satisfies (2) in T, then

$$\int_{\Gamma(T)} u_h^2\, \alpha\cdot n\, d\tau \le C\, \{\, h^{1/2}\, \|f\|_T^2 + \|P_{n-2}f\|_T^2 + h^{3/2}\, \|\nabla u_h\|_T^2 + \|u_h\|_T^2 \,\}. \tag{4}$$

Proof: Omitting the subscript T on the norms and inner products which follow, we integrate by parts to obtain

$$\begin{aligned}\frac{1}{2}\int_{\Gamma} u_h^2\, \alpha\cdot n\, d\tau &= (\,(u_h)_\alpha, u_h) = (\,(u_h)_\alpha, P_{n-1}u_h\,) \\ &= -\,(\,f, (I-P_{n-1})u_h\,) + (\,f, (I-P_{n-2})u_h\,) \\ &\quad + (\,f, P_{n-2}u_h\,) - (\,\beta u_h, P_{n-1}u_h\,).\end{aligned}$$

In the last of these equalities, equation (2) and the fact that $(u_h)_\alpha \in \mathbf{P}_{n-1}(T)$ were used. Applying standard estimates we obtain (4).

Lemma 5.2: If T is a type II triangle and u_h satisfies (3) on T, then for any $\epsilon > 0$

$$\begin{aligned}\int_{\Gamma(T)} u_h^2\, \alpha\cdot n\, d\tau &\le \epsilon\, h^{3/2}\, |(u_h)_\alpha|^2_{\Gamma_{out}(T)} \\ &\quad + C\, \{\, \|P_{n-2}f\|_T^2 + h^{3/2}\, \|\nabla u_h\|_T^2 + \|u_h\|_T^2 \,\}\end{aligned} \tag{5}$$

where C depends on ϵ.

Proof: Again omitting the subscript T and integrating by parts, we obtain

$$\begin{aligned}\frac{1}{2}\int_{\Gamma} u_h^2\, \alpha\cdot n\, d\tau &= (\,(u_h)_\alpha, u_h\,) \\ &= (\,(u_h)_\alpha, (I-P_{n-2})u_h\,) + (P_{n-2}f, P_{n-2}u_h\,) - (\beta u_h, P_{n-2}u_h\,).\end{aligned}$$

It then follows by standard estimates that

$$\int_\Gamma u_h{}^2 \, \alpha\cdot n \, d\tau \;\le\; C \,\{\; \|u_h\|^2 + \|P_{n-2}f\|^2 + \epsilon\, h^{1/2} \, \|(u_h)_\alpha\|^2$$

$$+\epsilon^{-1} \, h^{3/2} \, \|\nabla u_h\|^2 \,\} \qquad (\text{for any } \epsilon > 0).$$

Next, by establishing the equivalence of norms and then scaling, we obtain

$$\|(u_h)_\alpha\|^2 \;\le\; C \,\{\, h\, |(u_h)_\alpha|^2_{\Gamma_{out}(T)} + \|P_{n-2}(u_h)_\alpha\|^2 \,\}. \tag{6}$$

Applying equation (3) we easily get

$$\|(u_h)_\alpha\|^2 \;\le\; C \,\{\, h\, |(u_h)_\alpha|^2_{\Gamma_{out}(T)} + \|P_{n-2}f\|^2 + \|u_h\|^2 \,\}.$$

Inserting this result in the previous inequality and replacing ϵ by $\frac{\epsilon}{C}$ establishes the lemma.

We now combine Lemmas 4.1, 4.2, 5.1 and 5.2 into a single local stability result.

Theorem 5.1: There exists a positive constant M such that for a triangle T of either type

$$\int_{\Gamma(T)} \Big[\frac{h^{3/2} \left(\frac{du_h}{d\tau}\right)^2}{\alpha\cdot n} + u_h{}^2 \, \alpha\cdot n \,\Big] \, d\tau \;-\; 2\, h^{3/2} \int_{\Gamma(T)} \beta \cot\theta \, u_h \, (u_h)_\tau \, d\tau$$

$$+ \; M h^{1/2} \, \|(u_h)_\alpha\|^2_T$$

$$\le \; C \,\{\; h^{1/2} \, \|f\|^2_T + h^{-1/2} \, \|P_{n-2}f\|^2_T + \int_{\Gamma_{in}(T)} [\, h^{3/2}(u_h)^2_\tau + u_h{}^2 \,] \, d\tau \,\}.$$

where θ is the angle defined in Lemma 4.2.

Proof: Since the proof is lengthy, let us just mention the key ideas. The details can be found in [4]. First choose $v_h = -\,2\, h^{3/2} \, (u_h)_{\tau_1\tau_2} / [(\alpha\cdot n_1)(\alpha\cdot n_2)]$ in (2) and (3), and use the identities of Lemmas 4.1 and 4.2. Note that this is a legitimate choice since $v_h \in \mathbf{P}_{n-2}(T)$. Now add to this result inequalities (4) or (5) depending on the number of inflow sides of T. The key point is that for a two-inflow-side triangle, the term involving $(u_h)^2_\alpha$ integrated over Γ_3 coming from Lemma 4.1 has a positive sign on the left of the resulting inequality. A similar term coming from Lemma 5.2 and appearing on the right of the inequality has a small coefficient and thus can be dealt with. In a one-inflow-side triangle, this type of term appears only on the right, but can be bounded in terms of f using equation (2).

Next observe that for a type I triangle T, equation (2) implies that

$$(u_h)_\alpha = P_{n-1}\{f - \beta u_h\}.$$

This allows us to gain control over $\|(u_h)_\alpha\|_T^2$. For a type II triangle T, we use (6) and note that the first term on the right has already been controlled. Since (3) implies that

$$P_{n-2}(u_h)_\alpha = P_{n-2}\{f - \beta u_h\},$$

the second term is also easily controlled.

The local results of Theorem 5.1 lead to global stability for u_h and $\frac{du_h}{d\tau}$ along interelement boundaries. The norm in which we obtain stability is a weighted sum of L^2 norms of u_h and its tangential derivative taken along *fronts* F_j, which describe the forward boundary of the solution after it has progressed through the first j layers. More specifically, we define

$$F_0 = \Gamma_{in}(\Omega)$$

$$F_j = F_{j-1} \cup \Gamma_{out}(S_j) - \Gamma_{in}(S_j), \quad j = 1, 2, \ldots .$$

The main result of this section will be the following stability theorem:

Theorem 5.2: If u_h is the solution of Problem $\mathbf{P_h}$, then for h sufficiently small

$$\int_{F_j} \{ \frac{h^{3/2} (\frac{du_h}{d\tau})^2}{|\alpha \cdot n|} + u_h^2 |\alpha \cdot n| \} d\tau + M h^{1/2} \|(u_h)_\alpha\|_{\Omega_j}^2$$

$$\leq C \{ h^{1/2} \|f\|_{\Omega_j}^2 + h^{-1/2} \|P_{n-2} f\|_{\Omega_j}^2 + \int_{F_0} \{ \frac{h^{3/2} (\frac{du_h}{d\tau})^2}{|\alpha \cdot n|} + u_h^2 |\alpha \cdot n| \} d\tau \}$$

where $\Omega_j = \cup_{k \leq j} S_k$ and M is a positive constant.

Proof: For any triangle T, we infer from Theorem 5.1 that

$$\int_{\Gamma_{out}(T)} \{ \frac{h^{3/2} (\frac{du_h}{d\tau})^2}{|\alpha \cdot n|} + u_h^2 |\alpha \cdot n| \} d\tau - 2 h^{3/2} \int_{\Gamma_{out}(T)} \beta \cot\theta \, u_h (u_h)_\tau \, d\tau$$

$$+ M h^{1/2} \|(u_h)_\alpha\|_T^2$$

$$\leq \int_{\Gamma_{in}(T)} \{ \frac{h^{3/2} (\frac{du_h}{d\tau})^2}{|\alpha \cdot n|} + u_h^2 |\alpha \cdot n| \} d\tau + 2 h^{3/2} \int_{\Gamma_{in}(T)} \beta \cot\theta \, u_h (u_h)_\tau \, d\tau$$

$$+ C \{ h^{1/2} \|f\|_T^2 + h^{-1/2} \|P_{n-2} f\|_T^2 + h \int_{\Gamma_{in}(T)} [h^{3/2} (u_h)_\tau^2 + u_h^2] d\tau \}.$$

Summing over all triangles $T \in S_j$ yields

$$p_j + q_j + M h^{1/2} \|(u_h)_\alpha\|^2_{S_j} \tag{7}$$

$$\leq \{1 + O(h)\} p_{j-1} + q_{j-1} + C \{ h^{1/2} \|f\|^2_{S_j} + h^{-1/2} \|P_{n-2}f\|^2_{S_j} \},$$

where

$$p_j \equiv \int_{F_j} \{ \frac{h^{3/2} (\frac{du_h}{d\tau})^2}{|\alpha \cdot n|} + u_h^2 |\alpha \cdot n| \} \, d\tau$$

and

$$q_j \equiv 2 h^{3/2} \int_{F_j} \beta \cot\theta \, u_h (u_h)_\tau \, d\tau,$$

with the convention that integrals over F_j are taken *left to right* (thus fixing the sign of $(u_h)_\tau$ in the definition of q_j).

After making some technical modifications to (7), it is possible to sum the resulting inequality. The cancellation of terms in adjoining layers yields

$$p_j + q_j + M h^{1/2} \|(u_h)_\alpha\|^2_{\Omega_j} \leq C \{ p_0 + q_0 \tag{8}$$

$$+ h^{1/2} \|f\|^2_{\Omega_j} + h^{-1/2} \|P_{n-2}f\|^2_{\Omega_j} \}.$$

One can then show that (8) remains valid with q_j and q_0 deleted. The result is Theorem 5.2.

As a consequence of Theorem 5.2 , we can establish stability for u_h, ∇u_h and $(u_h)_\alpha$ over Ω.

Theorem 5.3:

$$\|u_h\|^2_\Omega + h^{3/2} \|\nabla u_h\|^2_\Omega + h^{1/2} \|(u_h)_\alpha\|^2_\Omega$$

$$\leq C \{ h^{1/2} \|f\|^2_\Omega + h^{-1/2} \|P_{n-2}f\|^2_\Omega + h^{3/2} |\frac{du_h}{d\tau}|^2_{\Gamma_{in}(\Omega)} + |u_h|^2_{\Gamma_{in}(\Omega)} \}.$$

Proof: This is proved by first obtaining the local estimates

$$\|u_h\|_T \leq C \{ h^{1/2} |u_h|_{\Gamma_{in}(T)} + h \|f\|_T \}$$

and

$$\|\nabla u_h\|_T \leq C \{ h^{1/2} |\frac{du_h}{d\tau}|_{\Gamma_{in}(T)} + h^{1/2} |u_h|_{\Gamma_{in}(T)} + \|f\|_T \}$$

valid for solutions of (2) or (3), and then using Theorem 5.2.

6. Error Estimates

To obtain error estimates for the method, we define an interpolant $u_I \in S_h^n$ by the following conditions:

(i). $u_I(a_i) = u(a_i)$ for all triangle vertices a_i.

(ii). $\int_\Gamma (u_I - u)\, \tau^l \, d\tau = 0, \quad l = 0, 1, ..., n-2$ for all triangle sides Γ.

(iii). $\int_T (u_I - u)\, q \, dx\, dy = 0$ for all $q \in \mathbf{P}_{n-3}(T)$ and all triangles T.

It is straightforward to show (for example, using the techniques in [2], Chapter 3) that u_I has the following approximation properties:

$$\|u - u_I\|_{j,T} \leq C h^{n+1-j} \|u\|_{n+1,T}, \quad j = 0, 1 \tag{9}$$

and

$$|u - u_I|_{j,\Gamma(T)} \leq C h^{n+1/2-j} \|u\|_{n+1,T}, \quad j=0, 1. \tag{10}$$

Rewriting equations (2) and (3) in the form

$$((u_h - u_I)_\alpha + \beta(u_h - u_I), v_h)_T = ((u - u_I)_\alpha + \beta(u - u_I), v_h)_T,$$

we may apply Theorems 5.2 and 5.3 with u_h replaced by $u_h - u_I$ and f replaced by $r \equiv (u - u_I)_\alpha + \beta(u - u_I)$. From (9), it follows immediately that

$$\|r\|_{\Omega_j} \leq C h^n \|u\|_{n+1,\Omega_j}.$$

Moreover, for all $v_h \in \mathbf{P}_{n-2}(T)$,

$$((u - u_I)_\alpha, v_h)_T = \int_\Gamma (u - u_I)\, v_h\, \alpha \cdot n \, d\Gamma - ((u - u_I), (v_h)_\alpha)_T = 0$$

since $(v_h)_\alpha \in \mathbf{P}_{n-3}(T)$. Hence

$$\|P_{n-2} r\|_{\Omega_j} \leq \|\beta(u - u_I)\|_{\Omega_j}$$

$$\leq C h^{n+1} \|u\|_{n+1,\Omega_j}.$$

We also assume, for convenience, that $u_h = u_I$ on $\Gamma_{in}(\Omega)$. Insertion of these bounds into Theorems 5.2 and 5.3 now yields:

Theorem 6.1: Let u be the solution of (1) and u_h the solution of Problem $\mathbf{P}_h$. If $u \in H^{n+1}(\Omega)$, there exists a constant C independent of h such that

$$\|u - u_h\|_\Omega \leq C h^{n+1/4} \|u\|_{n+1,\Omega}$$

$$\|\nabla(u - u_h)\|_\Omega \leq C h^{n-1/2} \|u\|_{n+1,\Omega}$$

$$\|(u - u_h)_\alpha\| \leq C h^n \|u\|_{n+1,\Omega_j},$$

and for $m = 1, 2, \ldots$

$$\{ \int_{F_j} (u - u_h)^2 \, d\tau \}^{1/2} \leq C h^{n+1/4} \|u\|_{n+1,\Omega_j}$$

and

$$\{ \int_{F_j} [\frac{d}{d\tau}(u - u_h)]^2 \, d\tau \}^{1/2} \leq C h^{n-1/2} \|u\|_{n+1,\Omega_j}.$$

REFERENCES

[1]. G. A. Baker, *A Finite Element Method for First Order Hyperbolic Equations*, Math. Comp., v. 29, 1975, pp. 995-1006.

[2]. P. G. Ciarlet, *The Finite Element Method for Elliptic Problems*, North-Holland, 1978.

[3]. T. Dupont, *Galerkin Methods for Modelling Gas Pipelines*, in Constructive and Computational Methods for Differential and Integral Equations, Lecture Notes in Math., v. 430, Springer-Verlag, 1974.

[4]. R. S. Falk and G. R. Richter, *Analysis of a Continuous Finite Element Scheme for Hyperbolic Equations*, preprint.

[5]. C. Johnson and J. Pitkaranta, *An Analysis of the Discontinuous Galerkin Method*, preprint.

[6]. P. Lesaint and P. A. Raviart, *On a Finite Element Method for Solving the Neutron Transport Equation*, in Mathematical Aspects of Finite Elements in Partial Differential Equations, C. deBoor, ed., Academic Press, 1974, pp. 89-123.

[7]. W. H. Reed and T. R. Hill, *Triangular Mesh Methods for the Neutron Transport Equation*, Los Alamos Scientific Laboratory Report LA-UR-73-479.

[8]. R. Winther, *A Stable Finite Element Method for First-Order Hyperbolic Systems*, Math. Comp., v. 36, 1981, pp. 65-86.

AN EFFICIENT MODULAR ALGORITHM FOR COUPLED NONLINEAR SYSTEMS

Tony F. Chan
Department of Computer Science
Yale University
New Haven, Connecticut 06520

Abstract

We present an efficient modular algorithm for solving the coupled system $G(u,t)=0$ and $N(u,t)=0$ where $u \in R^n$, $t \in R^m$, $G: R^n \times R^m \rightarrow R^n$ and $N: R^n \times R^m \rightarrow R^m$. The algorithm is modular in the sense that it only makes use of the basic iteration S of a general solver for the equation $G(u,t)=0$ *with t fixed*. It is therefore well-suited for problems for which such a solver already exists or can be implemented more efficiently than a solver for the coupled system. Local convergence results are given. Basically, if S is sufficiently contractive for G, then convergence for the coupled system is guaranteed. The algorithm is applied to two applications: (1) numerical continuation methods and (2) constrained optimization. Numerical results are given for the case where G represents a nonlinear elliptic operator. Three choices of S are considered: Newton's method, a two-level nonlinear multi-grid solver and a supported Picard iteration.

This work was supported in part by the Department of Energy under contract DE-AC02-81ER10996, and by the Army Reseach Office under contract DAAG-83-0177.

Keywords: *Coupled nonlinear systems, Newton's method, continuation methods, constrained optimization, multi-grid methods, supported Picard iteration.*

1. Introduction

This paper is concerned with numerical algorithms for solving coupled nonlinear systems of the form:

$$C(z) \equiv \begin{pmatrix} G(u,t) \\ N(u,t) \end{pmatrix} = 0$$

where $u \in R^n$, $t \in R^m$, $G : R^n \times R^m \mapsto R^n$ and $N : R^n \times R^m \mapsto R^m$. We assume that a solution z^* exists and that the Jacobian

$$J = \begin{pmatrix} G_u & G_t \\ N_u & N_t \end{pmatrix}$$

is nonsingular at z^*. We consider two applications in this paper. One is to continuation methods where G represents a nonlinear system in u with dependence on some parameters t and N represents an arclength condition constructed to follow the solution manifolds. Another application is to constrained optimization problems, where G represents the Lagrangian function, t the Lagrange multipliers and N the constraints.

In this paper, we are primarily interested in the case where G is large, sparse and structured, such as discretizations of partial differential equations. Very often, efficient techniques exist for exploiting such structures when one solves $G(u,t) = 0$ for u with t fixed. If $m \ll n$, it is desirable to be able to solve $C(z) = 0$ with about the same efficiency. However, while many conventional iterative algorithms can be applied directly to $C(z) = 0$ to solve for z^*, they often fail to exploit structures in G. Consider the use of Newton's method, for example. The Jacobian J of $C(z)$ does not necessarily inherit the following properties which the Jacobian G_u of G may possess: symmetry, positive definiteness, separability and bandedness [4].

In this paper, we present an algorithm for solving the coupled system which makes use of a general solver for $G = 0$, *for fixed t*. We assume that this solver is available in the form of a fixed point iteration operator S, which takes an approximate solution u_i and produce the next iterate $u_{i+1} = S(u_i, t)$. For example, for Newton's method we have $S^{Newton} = u - G_u^{-1}(u,t)G(u,t)$. The algorithm is not restricted to Newton's method, however, and in general any sufficiently convergent solver for G can be used. Such an algorithm can therefore exploit special efficient solvers specifically designed for solving $G = 0$.

In Section 2, we present the algorithm and some convergence results. In Section 3, we discuss applications to numerical continuation methods and constrained optimization. In Section 4, numerical results are presented for the case where G is an elliptic operator and for three choices of S: Newton's method, a nonlinear multi-grid method and a supported Picard iteration.

2. The Algorithm

Our algorithm is motivated by Newton's method applied to $C(z) = 0$. At each iteration, the following linear system

$$\begin{pmatrix} G_u & G_t \\ N_u & N_t \end{pmatrix} \begin{pmatrix} \delta u \\ \delta t \end{pmatrix} = -\begin{pmatrix} G \\ N \end{pmatrix}$$

has to be solved, often by the following block Gaussian elimination algorithm.

Algorithm BE:

1. Solve $G_u w = -G$ for w, where $w \in R^n$.
2. Solve $G_u v = G_t$ for v, where $v \in R^n \times R^m$.
3. Solve $(N_t - N_u v)\delta t = -(N + N_u w)$ for δt.
4. Compute $\delta u = w - v\delta t$.

The idea in the new algorithm is to use S to approximately solve for w and v in Steps 1 and 2 in Algorithm BE. Since the vector w is precisely the change in the iterate u in one step of Newton's method applied to $G(u,t) = 0$, it seems natural to approximate w by $S(u_i, t_i) - u_i$, where u_i and t_i are the current iterates. The situation for approximating v is slightly more complicated since it does not directly correspond to an iteration based on $G(u,t) = 0$. In [3], it is argued that it is reasonable to approximate v by $-S_t$. In particular, if $S = S^{Newton}$ then this approximation is exact. In practice, we can approximate S_t by a difference approximation. We therefore arrive at the following :

Algorithm ANM (*Approximate Newton Method*) : Given an initial guess (u_0, t_0), iterate the following steps until convergence:

1. Compute $w = S(u_i, t_i) - u_i$.
2. For $j = 1, m$ compute
$$v_j = -\frac{S(u_i, t_i + \epsilon_j e_j) - S(u_i, t_i)}{\epsilon_j},$$
where v_j denotes the j-th column of v, ϵ_j denotes a small finite difference interval and e_j denotes the j-th unit vector.
3. Solve the following m by m system for d:
$$(N_t(u_i, t_i) - N_u(u_i, t_i)v)d = -(N(u_i, t_i) + N_u(u_i, t_i)w)$$
4. Compute $t_{i+1} = t_i + d$.
5. Compute $u_{i+1} = u_i + w - vd$.

In [3], it is shown that if S is sufficient contractive, then the matrix in Step 3 of Algorithm ANM is nonsingular and thus the algorithm is well-defined.

The convergence of Algorithm ANM is analyzed in [3]. If we ignore the truncation error in the finite difference approximation to v in Step 2 of the algorithm, we have the following local convergence result:

Theorem 2.1. *Algorithm ANM converges (locally) iff $\rho(PS_u) < 1$, where $P \equiv I + v(N_t - N_u v)^{-1} N_u$.*

The following sufficient conditions for convergence follows immediately:

Theorem 2.2. *Algorithm ANM converges if $||S_u|| < \frac{1}{||P||}$, in any vector induced norm.*

Theorem 2.3. *If S_u is normal, then Algorithm ANM converges if $\rho(S_u) < \frac{1}{||P||_2}$. If P and S_u are both normal, or are simultaneously diagonalizable, then Algorithm ANM converges if $\rho(S_u) < \frac{1}{\rho(P)}$.*

If for a particular S, S_u does not satisfy any of these bounds, then we can define a modified iteration operator :
$$\hat{S}(u,t) = \overbrace{S(S \cdots S(S}^{k \text{ times}}(u,t), t), \cdots, t), t),$$
i.e. iterate S k times. Since $\hat{S}_u = S_u^k$, we can choose k large enough to make $\rho(\hat{S})$ or $||\hat{S}_u||$ small enough for convergence.

3. Applications

3.1. Continuation Methods

Here G represents a system of parameterized nonlinear equations, with u playing the role of the main variable and t the parameters. The goal is to trace solution manifolds of $G = 0$. This is often done by freezing all parameters except one (and therefore we restrict our attention to the case $m = 1$) and curve-following continuation methods [1, 5, 7, 8] are used to trace the branches corresponding to this one parameter. Usually, a predictor-corrector method is used. A predicted value is generated from a known solution (u, t) and the local unit tangent $(\dot{u}, \dot{t})$ defined by:
$$G_u \dot{u} + G_t \dot{t} = 0$$
$$||\dot{u}||_2^2 + \dot{t}^2 = 1.$$

The corrector is defined to be the solution of a coupled nonlinear system:
$$G(u(\delta s), \lambda(\delta s)) = 0$$
$$N(u(\delta s), \lambda(\delta s)) = 0,$$

where N defines a local parameterization of the solution curve with parameter δs. Two typical N's that are widely used in the literature are:

$$N^1 = \dot{u}_0^T(u-u_0) + \dot{t}_0(t-t_0) - \delta s,$$
$$N^2 = e_j^T\begin{pmatrix} u-u_0 \\ t-t_0 \end{pmatrix} - \delta s, \qquad 1 \le j \le n+1,$$

where (u_0, t_0) is a known solution on the solution curve, δs is a continuation step and e_j is the j-th unit vector. For more details the reader is referred to [7] for N^1 and [8] for N^2.

Algorithm ANM is especially well-suited for this application. First, for large problems, the solution of the coupled nonlinear system often constitutes the most costly part of the overall continuation process and Algorithm ANM does this efficiently by making it possible to exploit structures in G. A second advantage is that Algorithm ANM allows the continuation procedure itself (such as the step length control, the predictor, the tangent computation) to be separate from the specific solver for G, making it much easier for general purpose continuation codes [9] to be applied efficiently to application areas with specialized solvers (e.g. the Navier-Stokes equations).

Since the functions N^1 and N^2 are related to G itself, the convergence results of the last section can be refined further [3].

Theorem 3.1. *For N^1, as $\delta s \to 0$, Algorithm ANM converges locally if any one of the following conditions holds:*

1. $\|S_u\|_p < \frac{1}{2}$, *for $p = 2$ or ∞.*
2. $\|S_u\|_2 < 1$, *if P is normal.*
3. $\rho(S_u) < \frac{1}{2}$, *if S_u is normal.*
4. $\rho(S_u) < 1$, *if P and S_u are either both normal or simultaneously diagonalizable.*

Theorem 3.2. *For N^2, assuming that the index j is chosen such that $|(v)_j| = \max_{1\le i\le n} |(v)_i|$, Algorithm ANM converges if any one of the following conditions holds:*

1. $\|S_u\|_\infty < \frac{1}{2}$.
2. $\|S_u\|_2 < \frac{1}{1+\sqrt{n}}$.
3. $\|S_u\|_2 < 1$, *if P is normal.*
4. $\rho(S_u) < \frac{1}{1+\sqrt{n}}$, *if S_u is normal.*
5. $\rho(S_u) < 1$, *if P and S_u are either both normal or simultaneously diagonalizable.*

These sufficient conditions are very conservative in general and it is argued in [3] that in practice Algorithm ANM is convergent *whenever S is convergent for G.*

3.2. Constrained Optimatization

Another application of Algorithm ANM can be found in equality constrained optimization. Consider the problem:

$$\min F(u)$$
$$\text{subject to} \quad N_i(u) = 0 \qquad i = 1, \ldots, m.$$

Define the Lagrangian $L(u,t) \equiv F(u) + \sum_{i=1}^m t_i N_i(u)$. The first order condition for a minimum is:

$$G(u,t) \equiv \nabla L(u,t) = \nabla F(u) + \sum_{i=1}^m t_i \nabla N_i(u) = 0$$

and

$$N(u,t) \equiv \begin{Bmatrix} N_1(u) \\ \vdots \\ N_m(u) \end{Bmatrix} = 0,$$

which is in the form of a coupled nonlinear system. Algorithm ANM is well-suited for this problem if $m \ll n$ (i.e. relatively few constraints) and an efficient method is available for solving the unconstrained problem $\nabla F(u) = 0$. For linear constraints, the second term in G is constant and such a solver can easily be adapted to define an efficient S for solving $G(u,t) = 0$ for fixed t. Nonlinear constraints are more difficult to handle and we shall not dwell on that here. Thus, Algorithm ANM allows the addition of constraints to an unconstrained problem to be treated in a very efficient manner.

4. Numerical Results

4.1. Continuation Methods

We have applied Algorithm ANM to solve the following parameterized nonlinear equation:

$$u_{xx} + te^u = 0, \qquad 0 \le x \le 1 \tag{4.1}$$

with the boundary conditions

$$u(0) = u(1) = 0,$$

by the pseudo arclength continuation method using the parameterization N^1. This problem has one simple turning point at $(||u||_\infty \approx 1.3, t \approx 3.5)$, separating two branches of solution. Problem (4.1) is discretized on a uniform mesh with n interior grid points by a standard second order centered difference approximation. The resulting discrete system of nonlinear equations $G(u,t) = 0$ has dimension n.

We present numerical results for three solvers S. The first is simply S^{Newton}. The second is a two level full-approximation-scheme multi-grid method for solving (4.1) [2]. Briefly, two nonlinear Gauss-Seidel smoothing sweeps are used before the correction problem is injected onto and solved on a coarser grid (one with $\frac{n-1}{2}$ grid points, n odd), after which the correction is interpolated (linearly) and added to the solution on the original grid which is then smoothed again with two more smoothing sweeps. In our implementation, the coarse grid problem is solved by 4 iterations of Newton's method. The two-level multi-grid algorithm is representative of many multi-grid solvers in terms of convergence properties and extends to higher dimensional problems in a straightforward way. The third solver implements a supported Picard iteration [6]. The system $G = 0$ for (4.1) can be written as $Au = F(u,t)$, which naturally suggests the following Picard iteration:

$$u^{i+1} \leftarrow A^{-1}F(u^i,t).$$

Structures in A can be exploited, for example, a fast elliptic solver can be used for A^{-1}. Unfortunately, this iteration is convergent if and only if $\rho(A^{-1}F_u) < 1$ and this does not necessarily hold for a given problem. In many applications, however, there are only a few divergent eigenvalues of $A^{-1}F_u$. The main idea in the supported Picard iteration is to apply Newton's method in the subspace spanned by the eigenvectors of $A^{-1}F_u$ corresponding to eigenvalues with magnitude exceeding 1. This can be implemented in an efficient manner because the dimension of this subspace is small. This method is ideally suited for our problem because there is only one divergent eigenvalue on the upper solution branch and therefore the Newton's method is effectively applied to a scalar problem. For more details on the implementation, the reader is referred to [6].

To illustrate the convergence behavior of Algorithm ANM on this problem, we have chosen to show the iterations for two points on the solution branch, one on the lower branch and the other on the upper branch, both with $t_0 = 3$. An initial guess to the solution is obtained by an Euler predictor based on the unit tangent $(\dot{u}_0, \dot{t}_0)$ and the finite difference interval ϵ is chosen to be 0.0001. The iteration is stopped when $\max\{||\delta z||_\infty, ||G||_\infty, |N|\} \le 10^{-5}$. The results for the case $n = 31$ are tabulated in Tables 1 and 2. It can be seen that the convergence is quite satisfactory for all the solvers.

4.2. Constrained Optimization

We have also applied Algorithm ANM to the following minimization problem:

$$\textit{minimize} \qquad F(u) \equiv u^T A u + \lambda \sum_{i=1}^{n} e^{u_i}$$

$$\textit{subject to} \qquad \sum_{i=1}^{n} l_i u_i = 0$$

where A is the discrete Laplacian in one dimension with zero Dirichlet boundary condition.

Note that the first order condition for the unconstrained problem is exactly the discretization of (4.1) and since the constraint is linear, any one of the three solvers considered so far can easily be adapted to construct a solver for $\nabla L(u,t) = 0$ for fixed multipliers t.

We shall present numerical results for the case $\lambda = 3$ and $n = 31$ which corresponds to the results given for the continuation problem. Only the results for S^{Newton} and S^{Picard} will be shown. The unconstrained solution is as shown in Figure 1. The iterations for three different constraints are shown in Tables 3 - 5 and the corresponding solutions are shown in Figures 2-4. The convergence is rapid for all cases.

iter	$\|\|u\|\|_\infty$	t	$\|\|\delta z\|\|_\infty$	$\|\|G\|\|_\infty$	$\|N\|$
Newton					
0	0.7283397E+00	0.3190261E+01	0.0E+00	0.5E−04	0.7E−07
1	0.7308150E+00	0.3173259E+01	0.2E−01	0.2E−04	0.3E−07
2	0.7308277E+00	0.3173151E+01	0.1E−03	0.2E−06	0.5E−07
3	0.7308277E+00	0.3173151E+01	0.1E−06	0.1E−06	0.4E−07
Multi-Grid					
0	0.7283397E+00	0.3190261E+01	0.0E+00	0.5E−04	0.7E−07
1	0.7308369E+00	0.3173014E+01	0.2E−01	0.2E−04	0.2E−07
2	0.7307944E+00	0.3173375E+01	0.4E−03	0.2E−05	0.7E−07
3	0.7308255E+00	0.3173168E+01	0.2E−03	0.3E−06	0.1E−07
4	0.7308269E+00	0.3173157E+01	0.1E−04	0.1E−06	0.3E−07
5	0.7308279E+00	0.3173148E+01	0.8E−05	0.9E−07	0.3E−07
Picard					
0	0.7283341E+00	0.3190261E+01	0.0E+00	0.5E−04	0.6E−07
1	0.7308021E+00	0.3173283E+01	0.2E−01	0.6E−05	0.7E−07
2	0.7308219E+00	0.3173152E+01	0.1E−03	0.2E−06	0.7E−07
3	0.7308219E+00	0.3173152E+01	0.3E−06	0.2E−06	0.3E−07

Table 1: $\|\|u_0\|\|_\infty = 0.641, t_0 = 3.0, \delta s = 0.4$

iter	$\|\|u\|\|_\infty$	t	$\|\|\delta z\|\|_\infty$	$\|\|G\|\|_\infty$	$\|N\|$
Newton					
0	0.2075063E+01	0.2895127E+01	0.0E+00	0.3E−04	0.7E−07
1	0.2075096E+01	0.2893029E+01	0.2E−02	0.4E−05	0.1E−06
2	0.2075096E+01	0.2893032E+01	0.3E−05	0.4E−06	0.1E−06
Multi-Grid					
0	0.2075063E+01	0.2895127E+01	0.0E+00	0.3E−04	0.7E−07
1	0.2075097E+01	0.2893288E+01	0.2E−02	0.5E−05	0.4E−07
2	0.2075093E+01	0.2892977E+01	0.3E−03	0.1E−05	0.2E−07
3	0.2075096E+01	0.2893027E+01	0.5E−04	0.3E−06	0.4E−07
4	0.2075096E+01	0.2893031E+01	0.4E−05	0.4E−06	0.2E−07
Picard					
0	0.2075073E+01	0.2895126E+01	0.0E+00	0.3E−04	0.5E−07
1	0.2075105E+01	0.2893017E+01	0.2E−02	0.4E−05	0.6E−07
2	0.2075107E+01	0.2893031E+01	0.1E−04	0.4E−06	0.5E−07
3	0.2075107E+01	0.2893031E+01	0.3E−06	0.6E−06	0.1E−06

Table 2: $\|\|u_0\|\|_\infty = 1.973, t_0 = 3.0, \delta s = 0.4$

iter	$\|\|u\|\|_\infty$	t	$\|\|\delta z\|\|_\infty$	$\|\|G\|\|_\infty$	$\|N\|$
Newton					
0	$0.9987573E-01$	$0.0000000E+00$	$0.0E+00$	$0.1E+00$	$0.7E-01$
1	$0.1170606E+00$	$-0.5526059E+02$	$0.6E+02$	$0.3E-01$	$0.7E-08$
2	$0.1020966E+00$	$-0.5137571E+02$	$0.4E+01$	$0.9E-05$	$0.0E+00$
3	$0.1021055E+00$	$-0.5138050E+02$	$0.5E-02$	$0.1E-07$	$0.0E+00$
4	$0.1021055E+00$	$-0.5138050E+02$	$0.9E-07$	$0.1E-07$	$0.0E+00$
Picard					
0	$0.9987573E-01$	$0.0000000E+00$	$0.0E+00$	$0.1E+00$	$0.7E-01$
1	$0.1368882E+00$	$-0.5531568E+02$	$0.6E+02$	$0.7E-01$	$0.7E-08$
2	$0.1020990E+00$	$-0.5137750E+02$	$0.4E+01$	$0.1E-04$	$0.0E+00$
3	$0.1021054E+00$	$-0.5138047E+02$	$0.3E-02$	$0.2E-07$	$0.0E+00$
4	$0.1021054E+00$	$-0.5138047E+02$	$0.4E-06$	$0.2E-07$	$0.0E+00$

Table 3: Constraint: $u(16) = 0.0$

iter	$\|\|u\|\|_\infty$	t	$\|\|\delta z\|\|_\infty$	$\|\|G\|\|_\infty$	$\|N\|$
Newton					
0	$0.9987573E-01$	$0.0000000E+00$	$0.0E+00$	$0.1E+00$	$0.2E-01$
1	$0.1418061E+00$	$-0.5032611E+02$	$0.5E+02$	$0.3E-01$	$0.4E-08$
2	$0.1315020E+00$	$-0.4681818E+02$	$0.4E+01$	$0.1E-04$	$0.0E+00$
3	$0.1315103E+00$	$-0.4682352E+02$	$0.5E-02$	$0.4E-07$	$0.0E+00$
4	$0.1315104E+00$	$-0.4682352E+02$	$0.1E-05$	$0.2E-07$	$0.0E+00$
Picard					
0	$0.9987573E-01$	$0.0000000E+00$	$0.0E+00$	$0.1E+00$	$0.2E-01$
1	$0.1501235E+00$	$-0.5038120E+02$	$0.5E+02$	$0.6E-01$	$0.4E-08$
2	$0.1315220E+00$	$-0.4682264E+02$	$0.4E+01$	$0.2E-04$	$0.0E+00$
3	$0.1315103E+00$	$-0.4682348E+02$	$0.8E-03$	$0.2E-07$	$0.0E+00$
4	$0.1315103E+00$	$-0.4682349E+02$	$0.7E-05$	$0.2E-07$	$0.0E+00$

Table 4: Constraint: $u(16) = 0.05$

iter	$\|\|u\|\|_\infty$	t	$\|\|\delta z\|\|_\infty$	$\|\|G\|\|_\infty$	$\|N\|$
Newton					
0	$0.9987573E-01$	$0.0000000E+00$	$0.0E+00$	$0.1E+00$	$0.3E+00$
1	$0.2130776E+00$	$-0.1571455E+02$	$0.2E+02$	$0.2E-01$	$0.0E+00$
2	$0.2104818E+00$	$-0.1573100E+02$	$0.2E-01$	$0.4E-06$	$0.0E+00$
3	$0.2104816E+00$	$-0.1573133E+02$	$0.3E-03$	$0.3E-07$	$0.1E-07$
4	$0.2104816E+00$	$-0.1573133E+02$	$0.2E-05$	$0.3E-07$	$0.1E-07$
Picard					
0	$0.9987573E-01$	$0.0000000E+00$	$0.0E+00$	$0.1E+00$	$0.3E+00$
1	$0.2200581E+00$	$-0.1502713E+02$	$0.2E+02$	$0.2E-01$	$0.1E-06$
2	$0.2104825E+00$	$-0.1573232E+02$	$0.7E+00$	$0.2E-04$	$0.0E+00$
3	$0.2104816E+00$	$-0.1573128E+02$	$0.1E-02$	$0.6E-07$	$0.0E+00$
4	$0.2104815E+00$	$-0.1573128E+02$	$0.2E-05$	$0.3E-07$	$0.0E+00$

Table 5: Constraint: $u(8) + u(16) + u(24) = 0.5$

Figure 1: Unconstrained Solution

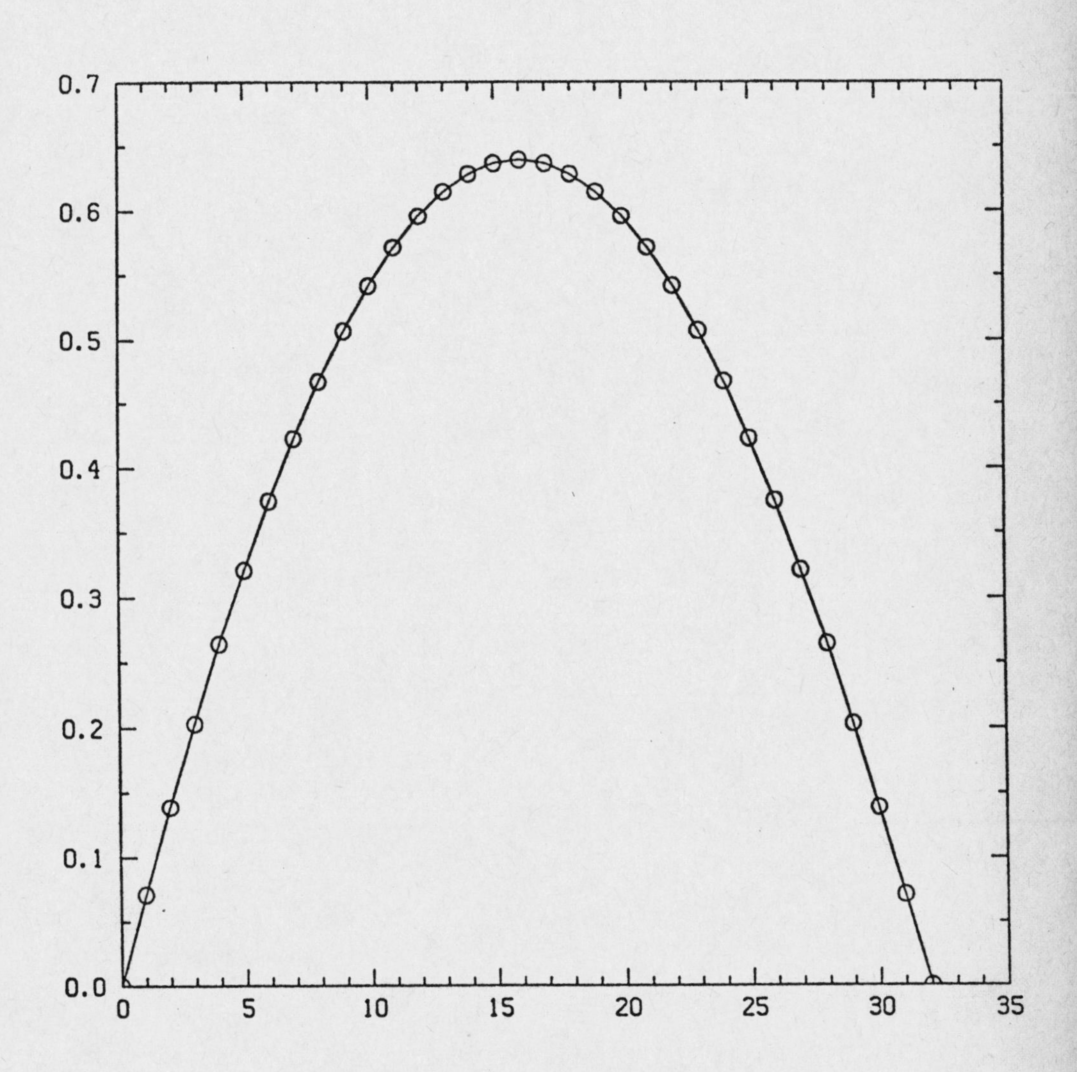

Figure 2: Constraint: $u(16) = 0.0$

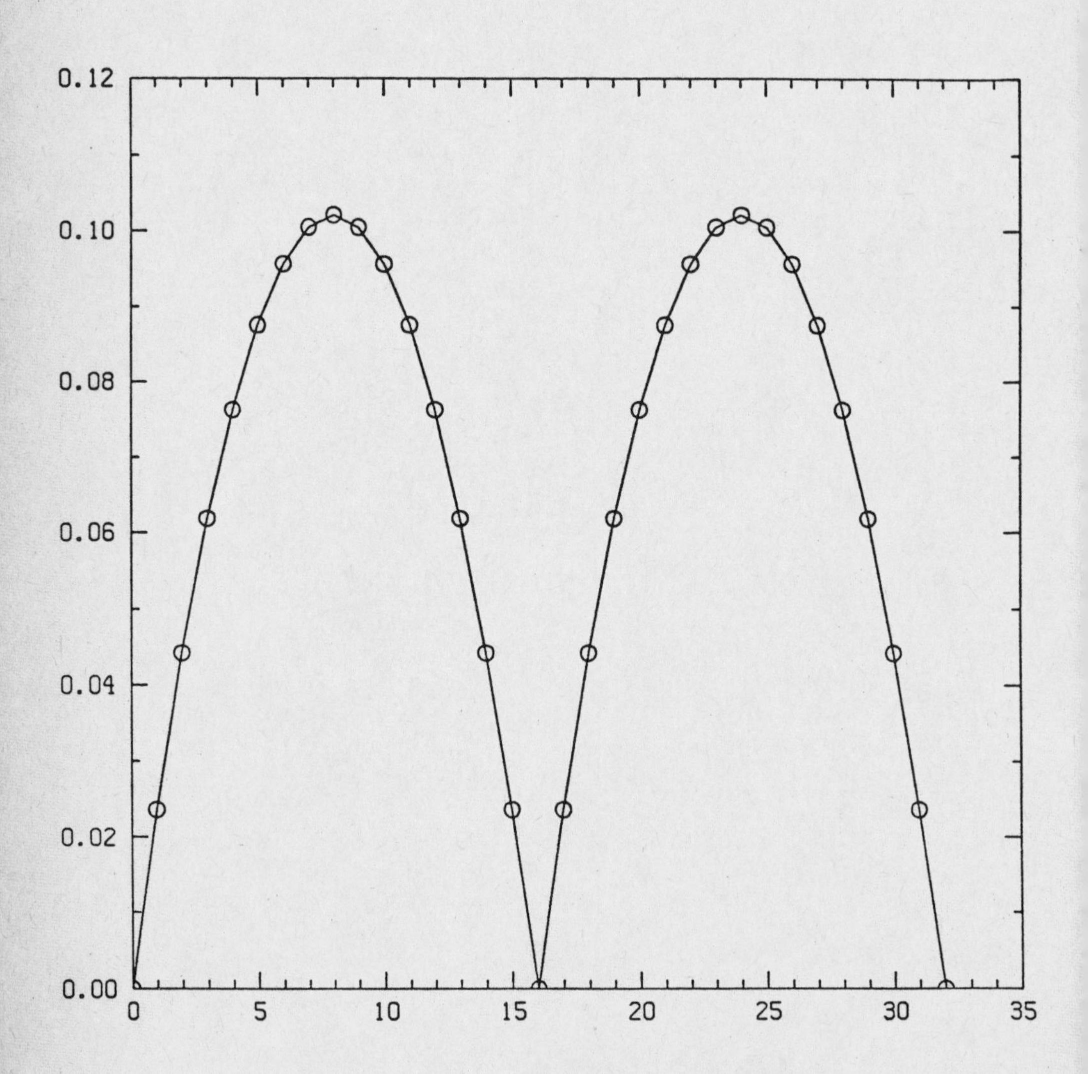

Figure 3: Constraint: $u(16) = 0.05$

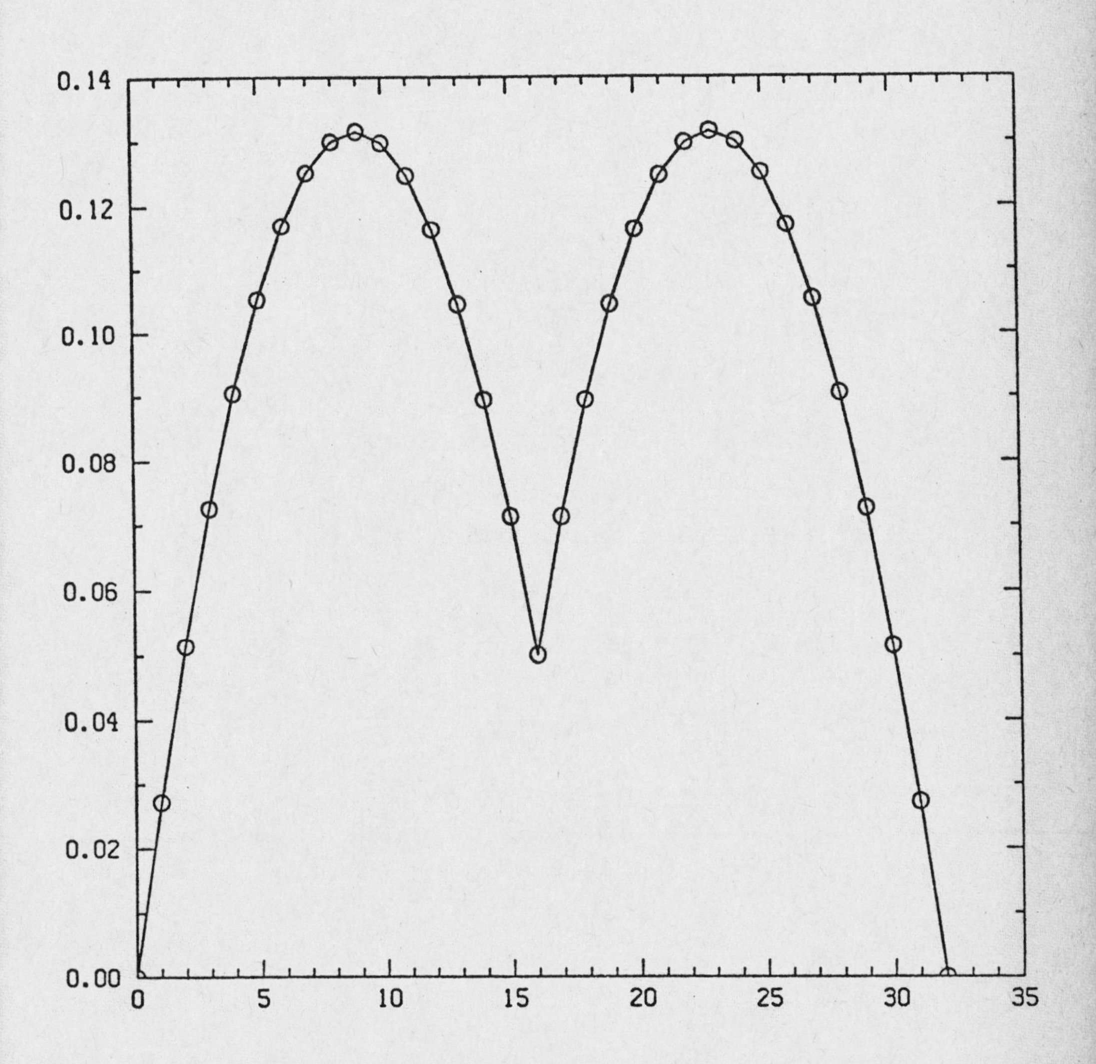

Figure 4: Constraint: $u(8) + u(16) + u(24) = 0.5$

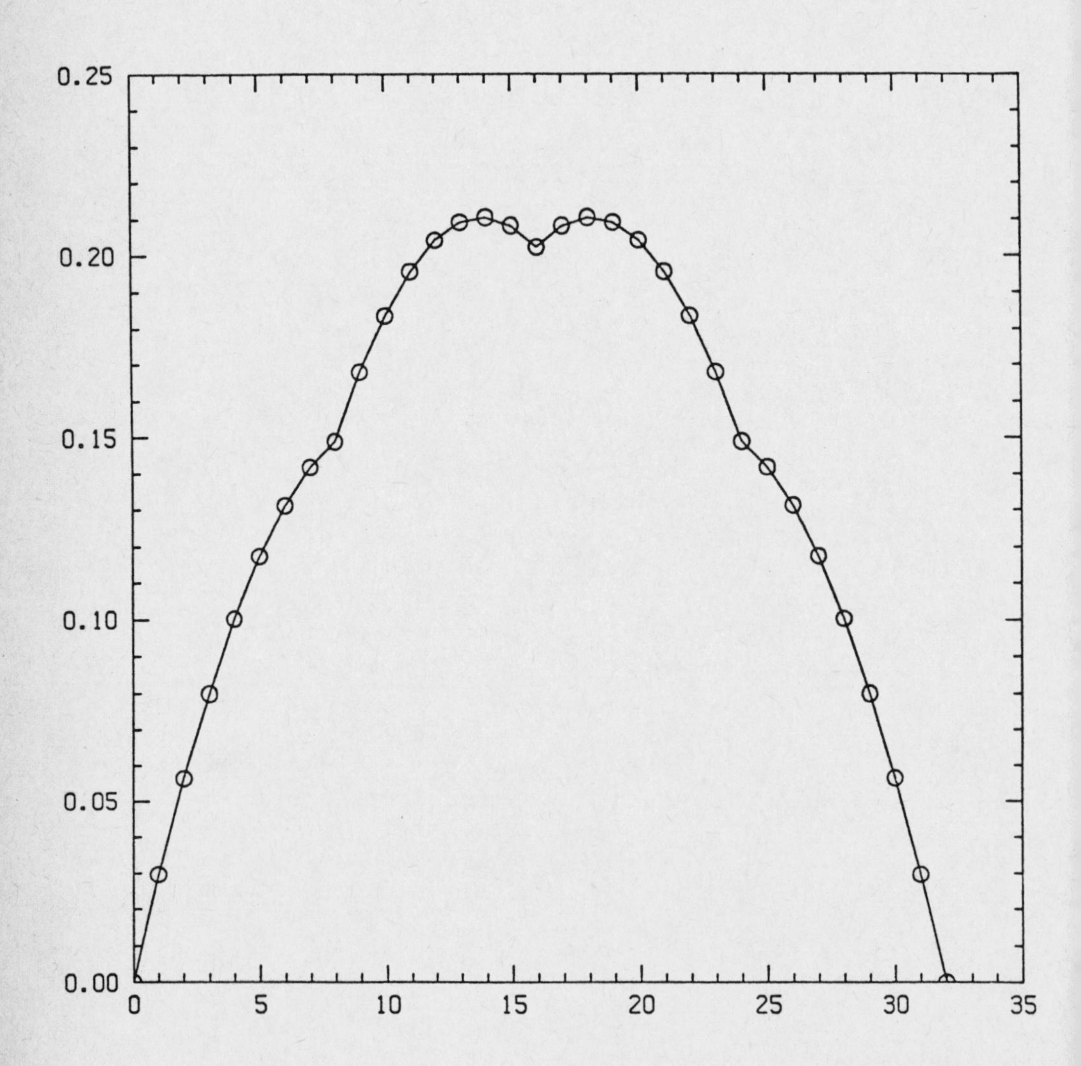

References

[1] E. Allgower and K. Georg, *Simplicial and Continuation Methods for Approximating Fixed Points and Solutions to Systems of Equations*, SIAM Review, 22/1 (1980), pp. 28–85.

[2] A. Brandt, *Multi-level Adaptive Solution to Boundary Value Problems*, Math. Comp., 31 (1977), pp. 333–390.

[3] T.F. Chan, *An Approximate Newton Method for Coupled Nonlinear Systems*, Technical Report Research Report YALEU/DCS/RR-300, Computer Science Department, Yale Univ., February 1984. Submitted to Siam J. Numer. Anal..

[4] ———, Techniques for Large Sparse Systems Arising from Continuation Methods, T. Kupper, H. Mittelmann and H. Weber eds., *Numerical Methods for Bifurcation Problems*, International Series of Numerical Math., Vol. 70, Birkhauser Verlag, Basel, 1984, pp. 116–128.

[5] C.B. Garcia and W.I. Zangwill, *Pathways to Solutions, Fixed Points and Equilibria*, Prentice-Hall, Englewood Cliffs, N.J., 1981.

[6] H. Jarausch and W. Mackens, *CNSP A Fast, Globally Convergent Scheme to Compute Stationary Points of Elliptic Variational Problems*, Technical Report 15, Institute for Geometrie und Praktische Mathematik, Rheinisch-Westfalische Technische Hochschule, Aachen, 1982.

[7] H.B. Keller, Numerical Solution of Bifurcation and Nonlinear Eigenvalue Problems, P. Rabinowitz ed., *Applications of Bifurcation Theory*, Academic Press, New York, 1977, pages 359–384.

[8] W.C. Rheinboldt, *Solution Fields of Nonlinear Equations and Continuation Methods*, SIAM J. Numer. Anal., 17 (1980), pp. 221–237.

[9] W.C. Rheinboldt and J.V.Burkardt, *A Locally Parameterized Continuation Process*, ACM Trans. Math. Soft., 9/2 June(1983), pp. 215–235.

OPTIMIZATION OF MULTISTAGE PROCESSES DESCRIBED BY DIFFERENTIAL-ALGEBRAIC EQUATIONS

K.R. Morison and R.W.H. Sargent
Imperial College, London.

Abstract

The paper describes an algorithm for the computation of optimal design and control variables for a multistage process, each stage of which is described by a system of nonlinear differential-algebraic equations of the form:

$$f(t,\dot{x}(t),x(t),u(t),v) = 0$$

where t is the time, x(t) the state vector, $\dot{x}(t)$ its time derivative, u(t) the control vector, and v a vector of design parameters. The system may also be subject to end-point or interior-point constraints, and the switching times may be explicitly or implicitly defined. Methods of dealing with path constraints are also discussed.

1. Introduction

Classical optimal control theory is concerned with the optimal performance of systems whose behaviour is described by a set of ordinary differential equations, and there is now a wide literature on both the underlying theory and numerical algorithms for solving such problems. However many systems of practical interest are naturally described in terms of a mixed set of differential and algebraic equations of the form

$$\begin{aligned} \dot{x}(t) &= f(t,x(t),y(t),u(t)) \\ 0 &= g(t,x(t),y(t),u(t)) \end{aligned} \qquad (1)$$

where $t \in T \subset R$ is the time, $x(t) \in R^n$ is a vector of state variables, $\dot{x}(t)$ is its derivative with respect to time, $y(t) \in R^m$ is a vector of associated state variables, $u(t) \in R^q$ is a vector of control variables, and $f{:}R{\times}R^n{\times}R^m{\times}R^q \to R^n$, $g{:}R{\times}R^n{\times}R^m{\times}R^q \to R^m$. For example, the differential equations describing a chemical process typically contain physical properties, such as density or specific enthalpy, which are given as instantaneous algebraic functions of the pressure, temperature and composition of the relevant mixture.

In principle, mixed systems like (1) can be reformulated as a system of ordinary differential equations, but this process may destroy the structure of the original system which can often be exploited in the numerical solution, particularly for large problems. Moreover, the reformulation can be far from trivial, as shown by Gear and Petzold (1983) and Pantelides (1985). There is therefore an incentive to extend the methods developed for ordinary differential equation (ODE) systems to apply directly to mixed systems. Indeed, as we shall see, there is no further complication involved in considering general systems of differential-algebraic equations (DAE) of the form:

$$f(t,\dot{x}(t),x(t),u(t),v) = 0 \qquad (2)$$

where $x(t),u(t)$ are as defined above, $v \in R^p$ is a vector of adjustable design parameters, and now $f:R \times R^n \times R^n \times R^q \times R^p \rightarrow R^n$.

Linear systems of type (2) are known as "descriptor systems" in the control literature, and are written in the form

$$E\dot{x}(t) = Ax(t) + Bu(t) \qquad (3)$$

where E, A, B are matrices, with E in general a singular matrix. The simulation and control of such systems has been studied by many authors - see for example Cobb (1983).

Multistage systems are also frequently of interest, in which a sequence of processes is used to achieve some overall result. Examples are a multi-stage rocket designed to achieve a maximum altitude, or the production of pharmaceuticals using a sequence of chemical reaction and separation processes.

In this paper we therefore consider a general multistage process, consisting of a sequence of stages j = 1,2, ... s, each of which is described by a DAE system of the form (2):

$$f^j(t,\dot{x}^j(t),x^j(t),u^j(t),v) = 0, \; t \in [t_{j-1},t_j], \qquad (4)$$

with $[t_o,t_s] \subset T \subset R, x^j(t) \in X^j \subset R^{n_j}$, $u^j(t) \in U^j(v) \subset R^{q_j}$, $v \in V \subset R^p$, $f^j:G^j \rightarrow R^{n_j}$, where $G^j \subset T \times X^j \times X^j \times R^{q_j} \times R^p$. We shall assume that the sets $T,V,X^j,U^j(v)$, j=1,2, ... s are compact, and that for each j=1,2, ... s the functions $f^j(.)$ and $u^j(.)$ are continuously differentiable with respect to their arguments on $[t_{j-1},t_j]$. However, discontinuities in the controls can be accommodated by subdividing the stages so that jumps occur between "stages".

The initial time t_o is taken as fixed, but the stage end-times, t_j, j=1,2,...s are implicitly defined as the times at which the corresponding "switching functions":

$$h^j(t,\dot{x}^j(t),x^j(t),u^j(t),v) \geq 0, \; t \in [t_{j-1},t_j] \qquad (5)$$

are reduced to zero. Of course a specified end-time t_j^* can be accommodated by writing $h^j = t_j^* - t$, and t_j^* could be included as an element of v if it is adjustable; however, in the actual implementation of the algorithm it will be more efficient to cater explicitly for specified end-times.

The initial state will normally be subject to some equality constraints (in addition to (4)) and it is also necessary to specify "junction conditions" relating the states in successive stages:

$$J^o(x^1(t_o),v) = 0, \quad J^j(x^{j+1}(t_j),x^j(t_j),v) = 0, \quad j = 1,2, \dots (s-1) \quad (6)$$

We consider later the restrictions on the functions $J^j(.)$, $j=0,1,2, \dots (s-1)$. We also allow stage end-time constraints of the form:

$$a^j \leq F^j(t_j,\dot{x}^j(t_j),x^j(t_j),u^j(t_j),v) \leq b^j, \quad j = 0,1,2, \dots s \quad (7)$$

where the $F^j: G^j \to R^{r_j}$ are again continuously differentiable with respect to their arguments, and $a^j,b^j \in R^{r_j}$ are constant vectors. We note that equality constraints can be accommodated by making the corresponding elements of a^j, b^j equal.

The optimization problem is to choose design parameters, v, and controls $u^j(t)$, $t\varepsilon[t_{j-1},t_j]$, $j = 1,2, \dots s$, to minimize the objective function:

$$P(t_s,\dot{x}^s(t_s),x^s(t_s),u^s(t_s),v), \quad (8)$$

subject to satisfying (4), (5), (6) and (7).

Again we assume that $P: G^s \to R$ is continuously differentiable with respect to its arguments.

2. The Numerical Solution

To solve the optimization problem, we use the same approach as that used by Sargent and Sullivan (1977) for the classical problem. This involves choosing a set of basis functions, involving a finite number of parameters $w^j \in R^{l_j}$, to represent the controls in each interval:

$$u^j(t) = \theta^j(t,w^j), \quad t \in [t_{j-1},t_j], \quad j = 1,2, \dots s, \quad (9)$$

and again we can subdivide intervals to obtain an adequate approximation with the given set of basis functions.

The solution of system (4) is then completely defined by the set of decision variables:

$$z = \{v, w^1, w^2, \dots w^s\} \tag{10}$$

Given this set of values, the trajectory $x^j(t)$, $t \in [t_{j-1}, t_j]$, $j = 1,2, \dots s$, can be obtained by successively solving the DAE systems (4) over each sub-interval in turn, using the switching conditions (5) to determine the t_j and the junction conditions (6) for the states. The constraint functions (7) and objective function (8) can then be evaluated, defining these values as functions of the decision variables z:

$$\phi^j(z) = F^j(t_j, \dot{x}^j(t_j), x^j(t_j), u^j(t_j), v), \tag{11}$$

$$\psi(z) = P\,(t_s, \dot{x}^s(t_s), x^s(t_s), u^s(t_s), v). \tag{12}$$

We thus reduce the problem to the nonlinear programme:

$$\begin{aligned} &\text{Choose } z \text{ to minimize } \psi(z), \\ &\text{subject to: } a^j \leq \phi^j(z) \leq b^j, \quad j = 1,2, \dots s, \\ &\qquad \theta^j(t,w^j) \in U^j(v), \; t \in [t_{j-1},t_j], \quad j = 1,2, \dots s \\ &\qquad v \in V \end{aligned} \tag{13}$$

This is still an infinite-dimensional problem, since the control constraints must be satisfied at all times. However the admissible sets $U^j(v)$ are usually defined by inequalities (often linear), and the basis functions $\theta^j(\cdot,\cdot)$ can usually be chosen so that verification at a finite set of times suffices. For example we may have

$$\underline{u}^j \leq u^j(t) \leq \bar{u}^j, \quad t \in [t_{j-1},t_j],$$

and then the choice $\quad u^j(t) = w^j, \quad t \in [t_{j-1},t_j], \qquad (14)$

gives the simple bounds $\quad \underline{u}^j \leq w^j \leq \bar{u}^j, \quad j = 1,2, \dots s.$

The resulting finite-dimensional nonlinear programme can then be solved by well known techniques, for example one of the recursive quadratic programming algorithms described by Sargent (1981).

Such algorithms require derivatives of the objective and constraint functions with respect to the decision variables, and expressions for these are derived in Section 3 in terms of the adjoint system for (4).

The system and adjoint equations can be solved directly using a technique proposed by Gear (1971). Most integration formulae for solving ordinary differential equations can be written in the form:

$$x_k = \gamma_k h_k \dot{x}_k + \phi_{k-1} \tag{15}$$

where $h_k = t_k - t_{k-1}$, $y_k \approx y(t_k)$, $\dot{y}_k \approx \dot{y}(t_k)$ and ϕ_{k-1} depends on past values of t_k, y_k, $\dot{y}_k$. The parameter γ_k is non-zero for an implicit integration formula, and for such formulae (15) may be solved for $\dot{x}_k$ and then substituted into (4) to yield:

$$f^j(t_k,(\gamma_k h_k)^{-1}(x_k^j - \phi_{k-1}), x_k^j, u^j(t_k), v) = 0 \tag{16}$$

This is a set of nonlinear equations which can be solved for x_k^j (usually using Newton's method), whereupon values of $\dot{x}_k^j$ can be obtained from (15). The usual tests to determine step-length and the appropriate order of the integration formula can still be applied, as for ODE systems.

Since there is a system of nonlinear equations to be solved at each time step in any case, and the Jacobian matrix is required for the adjoint system, there is no penalty involved in using an implicit formula, designed for dealing with stiff systems. The most popular method in this class is Gear's BDF method, but the implicit Runge-Kutta methods (see Cameron, 1983, and Burrage, 1982) also deserve further investigation.

3. Expressions for Derivatives

We start by deriving the adjoint system for the linear DAE system:

$$A(t)\,\dot{x}(t) + B(t)x(t) = 0 \tag{17}$$

Introducing an integrating factor $\lambda(t)$, we write

$$\lambda(t)A(t)dx + \lambda(t)B(t)x(t)dt = 0.$$

For the left-hand side to be an exact differential we then require

$$\frac{\partial}{\partial t}(\lambda(t)A(t)) = \frac{\partial}{\partial x}(\lambda(t)B(t)x(t)), \tag{18}$$

which yields

$$\frac{d}{dt}(\lambda(t)A(t)) = \lambda(t)B(t)$$

This is the adjoint system, which is more conveniently written:

$$\dot{\mu}(t) = \lambda(t)B(t) \quad ; \quad \mu(t) = \lambda(t)A(t) \tag{19}$$

For the nonlinear system (4) we consider two sets of decision variables, z and $\bar{z}$, with corresponding solutions:

$$\{t, x^j(t), u^j(t), v\} \quad , \quad t \in [t_{j-1}, t_j]$$

$$\{t, \bar{x}^j(t), \bar{u}^j(t), \bar{v}\} \quad , \quad t \in [\bar{t}_{j-1}, \bar{t}_j]$$

and define $\delta z = \bar{z} - z \quad , \quad \delta v = \bar{v} - v, \ \delta t_j = \bar{t}_j - t_j \quad ,$

$$\delta x^j(t) = \bar{x}^j(t) - x^j(t), \quad \Delta x^j(t_j) = \bar{x}^j(\bar{t}_j) - x^j(t_j), \text{ etc.}$$

Without loss of generality, we can take $0 < \delta t_j < t_{j+1} - t_j$. Since $\bar{x}^{j+1}(t_j)$ and $x^j(\bar{t}_j)$ do not then exist, we have the relations:

$$\begin{aligned} \Delta x^j(t_j) &= \dot{\bar{x}}^j(t_j)\cdot\delta t_j + \delta x^j(t_j) + o(\delta t_j) \\ \Delta x^{j+1}(t_j) &= \delta x^{j+1}(\bar{t}_j) + \dot{x}^{j+1}(t_j)\delta t_j + o(\delta t_j) \end{aligned} \tag{20}$$

To simplify the derivation which follows, we also adopt the abbreviated notation:

$$f^j(t) = f^j(t, \dot{x}^j(t), x^j(t), u^j(t), v)$$

$$\bar{f}^j(t) = f^j(t, \dot{\bar{x}}^j(t), \bar{x}^j(t), \bar{u}^j(t), \bar{v})$$

$$\tilde{f}^j(t) = f^j(t, \dot{\bar{x}}(t), \bar{x}^j(t), u^j(t), v)$$

and similarly for other functions evaluated along the solution trajectory. We then define the adjoint system for equations linearized about the base trajectory:

$$\dot{\mu}^j(t) = \lambda^j(t) f_x^j(t) \quad , \quad \mu^j(t) = \lambda^j(t) f_{\dot{x}}^j(t), \quad j = 1,2, \dots s \qquad (21)$$

Using the Mean Value Theorem, we have

$$\tilde{f}^j(t) - f^j(t) = f_{\dot{x}}^j(t).\delta\dot{x}(t) + f_x^j(t)\delta x(t) + \xi^j(t), \qquad (22)$$

where $\xi^j(t) = o\{\|x^j(.)\|\}$, with the norm $\|.\|$ defined by

$$\|x^j(.)\| = \sup_{t \epsilon [\bar{t}_{j-1}, t_j]} \max\{|x(t)|, |\dot{x}(t)|\} \qquad (23)$$

where $|.|$ is the Euclidean vector norm.

Then from (21) and (22) we have

$$\frac{d}{dt}\{\mu^j(t)\delta x^j(t)\} = \dot{\mu}^j(t)\delta x^j(t) + \mu^j(t)\delta\dot{x}^j(t)$$

and
$$\mu^j(t_j)\delta x^j(t_j) = \mu^j(\bar{t}_{j-1})\delta x(\bar{t}_{j-1}) + \int_{\bar{t}_{j-1}}^{t_j} \lambda^j(t)\{\tilde{f}^j(t) - \xi(t)\}dt. \qquad (24)$$

We now form a Lagrangian function for the variation $\Delta P(t_s)$, with equality constraints over the subintervals and at junction points adjoined using appropriate multipliers:

$$\Delta\psi = \Delta P(t_s) + \sum_{j=0}^{s-1}\{\zeta^j \Delta J^j(t_j) + \omega^j \Delta f^{j+1}(t_j)\} + \sum_{j=1}^{s}\{\upsilon^j \Delta f^j(t_j) + \pi^j \Delta h^j(t_j)\}$$

$$+ \sum_{j=1}^{s}\{\int_{\bar{t}_j}^{t_j} \lambda^j(t)[\tilde{f}^j(t) - \xi^j(t)]dt + \mu^j(\bar{t}_{j-1})\delta x^j(\bar{t}_{j-1}) - \mu^j(t_j)\delta x^j(t_j)\} \qquad (25)$$

Again using the Mean Value Theorem, we can expand each of the function differences in (25) to first order, and after collecting terms this yields:

$$\Delta\psi = \sum_{j=0}^{s-1} \omega^j f_{\dot{x}}^{j+1}(t_j).\Delta\dot{x}^{j+1}(t_j) + \sum_{j=0}^{s-1}\{\zeta^j J^j_{x^{j+1}}(t_j) + \omega^j f_x^{j+1}(t_j)\}\Delta x^{j+1}(t_j)$$

$$+ \sum_{j=0}^{s-1} \omega^j f_u^{j+1}(t_j)\Delta u^{j+1}(t_j) + \sum_{j=0}^{s-1}\{\zeta^j J_v^j(t_j) + \omega^j f_v^{j+1}(t_j)\}\delta v + \sum_{j=0}^{s-1} \mu^{j+1}(t_j)\delta x^{j+1}(\bar{t}_j)$$

$$+\sum_{j=1}^{s-1}\{\omega^j f_t^{j+1}(t_j)+\upsilon^j f_t^j(t_j)+\pi^j h_t^j(t_j)\}\delta t_j+\sum_{j=1}^{s-1}\{\upsilon^j f_{\dot{x}}^j(t_j)+\pi^j h_{\dot{x}}^j(t_j)\}\Delta\dot{x}^j(t_j)$$

$$+\sum_{j=1}^{s-1}\{\zeta^j J_{x^j}^j(t_j)+\upsilon^j f_x^j(t_j)+\pi^j h_x^j(t_j)\}\Delta x^j(t_j)-\sum_{j=1}^{s-1}\mu^j(t_j)\delta x^j(t_j)$$

$$+\sum_{j=1}^{s-1}\{\upsilon^j f_u^j(t_j)+\pi^j h_u^j(t_j)\}\Delta u^j(t_j)+\sum_{j=1}^{s-1}\{\upsilon^j f_v^j(t_j)+\pi^j h_v^j(t_j)\}\delta v$$

$$+\{P_t(t_s)+\upsilon^s f_t^s(t_s)+\pi^s h_t^s(t_s)\}\delta t_s+\{P_{\dot{x}}(t_s)+\upsilon^s f_{\dot{x}}^s(t_s)+\pi^s h_{\dot{x}}^s(t_s)\}\Delta\dot{x}^s(t_s)$$

$$+\{P_x(t_s)+\upsilon^s f_{\dot{x}}^s(t_s)+\pi^s h_x^s(t_s)\}\Delta x^s(t_s)+\{P_u(t_s)+\upsilon^s f_u^s(t_s)+\pi^s h_u^s(t_s)\}\Delta u^s(t_s)$$

$$+\{P_v(t_s)+\upsilon^s f_v^s(t_s)+\pi^s h_v^s(t_s)\}\delta v - \mu^s(t_s)\delta x^s(t_s)$$

$$+\sum_{j=1}^{s}\int_{\bar{t}_{j-1}}^{t_j}\lambda^j(t)\{\tilde{f}^j(t)-\xi^j(t)\}dt + \sum_{j=1}^{s}\eta^j \tag{26}$$

where $\eta^j = o\{\delta t_j\} + o\{\|\delta x^j(.)\|\} + o\{\|\delta u^j(.)\|\} + o\{\delta v\}$, $j = 1,2, \dots s$.

We now substitute from (20) into (26), and then choose the multipliers so that terms in $\Delta\dot{x}^{j+1}(t_j)$, $\delta x^{j+1}(\bar{t}_j)$, $j = 0,1, \dots (s-1)$, and δt_j, $\Delta\dot{x}^j(t_j)$, $\delta x^j(t_j)$, $j=1,2,\dots s$ vanish, leaving the simplified expression:

$$\Delta\psi = \sum_{j=0}^{s-1}\omega^j f_u^{j+1}(t_j)\ \delta u^{j+1}(\bar{t}_j) + \sum_{j=0}^{s-1}\{\zeta^j J_v^j(t_j) + \omega^j f_v^{j+1}(t_j)\}\ \delta v$$

$$+ \sum_{j=1}^{s}\{\upsilon^j f_u^j(t_j) + \pi^j h_u^j(t_j)\}\ \delta u^j(t_j) + \sum_{j=1}^{s}\{\upsilon^j f_v^j(t_j) + \pi^j h_v^j(t_j)\}\ \delta v$$

$$+ P_u(t_s)\ \delta u^s(t_s)+P_v(t_s)\delta v+\sum_{j=1}^{s}\int_{\bar{t}_{j-1}}^{t_j}\lambda^j(t)\{\tilde{f}^j(t) - \xi^j(t)\}dt+\sum_{j=1}^{s}\eta_j \tag{27}$$

Provided that $\delta\dot{x}^j(t) \to 0$, $\delta x^j(t) \to 0$, $t\varepsilon\ [t_{j-1},t_j]$, and $\delta t_j \to 0$, $j=1,2,\dots s$, as

$\delta z \to 0$, it is now straightforward to obtain partial derivatives from (27) by taking all but one perturbation to be zero, dividing by the non-zero perturbation, and taking the limit as it tends to zero. This process yields:

$$\psi_v = \sum_{j=0}^{s-1} \{\zeta^j J_v^j(t_j) + \omega^j f_v^{j+1}(t_j)\} + \sum_{j=1}^{s} \{\upsilon^j f_v^j(t_j) + \pi^j h_v^j(t_j)\}$$

$$+ P_v(t_s) + \sum_{j=1}^{s} \int_{t_{j-1}}^{t_j} \lambda^j(t) f_v^j(t)\, dt \tag{28}$$

$$\psi_{w^j} = \omega^{j-1} f_u^j(t_{j-1})\theta^j_{w^j}(t_{j-1}) + \{\upsilon^j f_u^j(t_j) + \pi^j h_u^j(t_j)\}\ \theta^j_{w^j}(t_j)$$

$$+ P_u(t_s)\theta^s_{w^s}(t_s)\cdot\delta_{js} + \int_{t_{j-1}}^{t_j} \lambda^j(t) f_u^j(t)\ \theta^j_{w^j}(t)\, dt \tag{29}$$

The multipliers satisfy the following relations:

$$\omega^j f_{\dot{x}}^{j+1}(t_j) = 0 \qquad j = 0,1,..(s-1)$$

$$\zeta^j J^j_{x^{j+1}}(t_j) + \omega^j f_x^{j+1}(t_j) + \mu^{j+1}(t_j) = 0 \qquad j = 0,1,..(s-1)$$

$$\upsilon^j f_{\dot{x}}^j(t_j) + \pi^j h_{\dot{x}}^j(t_j) = 0 \qquad j = 1,2,..(s-1)$$

$$\zeta^j J^j_{x^j}(t_j) + \upsilon^j f_x^j(t_j) + \pi^j h_x^j(t_j) = \mu^j(t_j) \qquad j = 1,2,..(s-1)$$

$$\zeta^j \{J^j_{x^{j+1}}(t_j)\dot{x}^{j+1}(t_j) + J^j_{x^j}(t_j)\dot{x}^j(t_j)\} + \omega^j Df^{j+1}(t_j) +$$

$$+ \upsilon^j Df^j(t_j) + \pi^j Dh^j(t_j) = 0 \qquad j = 1,2,..(s-1)$$

$$P_{\dot{x}}(t_s) + \upsilon^s f_{\dot{x}}^s(t_s) + \pi^s h_{\dot{x}}^s(t_s) = 0$$

$$P_x(t_s) + \upsilon^s f_x^s(t_s) + \pi^s h_x^s(t_s) = \mu^s(t_s)$$

$$DP(t_s) + \upsilon^s Df^s(t_s) + \pi^s Dh^s(t_s) = 0, \tag{30}$$

where $$Df^j(t) = f_t^j(t) + f_x^j(t)\dot{x}(t) + f_u^j(t)\dot{u}(t), \text{ etc.} \tag{31}$$

The structure of the system is more evident if we partition the state vector $x^j(t)$ into the set of variables whose derivatives appear in (4) or (5), which we shall still denote by $x^j(t)$, and the remaining variables, which we shall denote by $y^j(t)$. Thus (4) becomes

$$f^j(t,\dot{x}^j(t),x^j(t),y^j(t),u^j(t),v) = 0, \quad t \in [t_{j-1},t_j], \tag{32}$$

and (5) becomes

$$h^j(t,\dot{x}^j(t),x^j(t),y^j(t),u^j(t),v) \geq 0, \quad t \in [t_{j-1},t_j], \tag{33}$$

It follows from (21) that the elements of $\mu^j(t)$ corresponding to $y^j(t)$ are zero, so we still write $\mu^j(t)$ for the non-zero portion corresponding to $x^j(t)$, and equation (21) can then be rewritten:

$$\dot{\mu}^j(t) = \lambda^j(t)f_x^j(t) \quad , \quad [\mu^j(t),0] = \lambda^j(t)[f_{\dot{x}}^j(t),f_y^j(t)]. \tag{34}$$

It follows from (30) that $P_{\dot{y}}(t_s)$ must be zero for consistency, and (30) can then be rewritten:

$$[\upsilon^s,\pi^s]\begin{bmatrix} Df^s(t_s), & f_{\dot{x}}^s(t_s), & f_y^s(t_s) \\ & & \\ Dh^s(t_s), & h_{\dot{x}}^s(t_s), & h_y^s(t_s) \end{bmatrix} = -[DP(t_s),\ P_{\dot{x}}(t_s),\ P_y(t_s)] \tag{a}$$

$$\mu^s(t_s) = P_x(t_s) + \upsilon^s f_x^s(t_s) + \pi^s h_x^s(t_s) \tag{b}$$

$$[\omega^j\ ,\ \zeta^j]\begin{bmatrix} f_{\dot{x}}^{j+1} & f_y^{j+1} & f_x^{j+1} \\ & & \\ 0 & J_{y^{j+1}}^j & J_{x^{j+1}}^j \end{bmatrix} = -[0\ ,\ 0\ ,\ \mu^{j+1}(t_j)],$$

$$j = 0,1, \ldots (s-1), \tag{c}$$

$$[\upsilon^j , \pi^j] \begin{bmatrix} Df^j(t_j) , & f^j_{\dot{x}}(t_j) , & f^j_y(t_j) \\ Dh^j(t_j) , & h^j_{\dot{x}}(t_j) , & h^j_y(t_j) \end{bmatrix} =$$

$$= -[\zeta^j \{J^j_{x^{j+1}}(t_j)\dot{x}^{j+1}(t_j) + J^j_{x^j}(t_j)\dot{x}^j(t_j)\} + \omega^j Df^{j+1}(t_j), 0, \zeta^j J^j_{y^j}(t_j)],$$

$$j=1,2,..(s-1), \quad (d)$$

$$\mu^j(t_j) = \zeta^j J^j_{x^j}(t_j) + \upsilon^j f^j_x(t_j) + \pi^j h^j_x(t_j), \; j = 1,2, \ldots (s-1). \quad (e) \qquad (35)$$

Thus, having chosen a set of decision variables z and integrated the system equations (32), using (33) and (6), to obtain the trajectory $[x^j(t), y^j(t)]$, $j=1,2,\ldots s$, we obtain the adjoint trajectory $[\mu^j(t), \lambda^j(t)]$ as follows:

1. set $j=s$, solve (35a) for υ^s, π^s, then compute $\mu^s(t_s)$ from (35b).

2. Integrate (34) from t_j to t_{j-1} to obtain $\mu^j(t), \lambda^j(t)$, $t \varepsilon [t_{j-1}, t_j]$.

3. Solve (35c) for $\omega^{j-1}, \zeta^{j-1}$.

4. If $j=1$, stop, otherwise set $j:=j-1$.

5. Solve (35d) for υ^j, π^j, then compute $\mu^j(t_j)$ from (35e).

6. Return to step 2.

Derivatives may then be evaluated from (28) and (29). A similar procedure may be used to generate corresponding adjoint systems and derivatives for any constraint function $F^j(.)$ in (7), where of course the intervals considered extend up to the stage end-time in question.

Having obtained function and derivative values in this way, an iteration of the nonlinear programming algorithm yields a new set of decision variables.

4. Existence and Regularity of Solutions

So far the development has been purely formal, and we have assumed existence, regularity and differentiability as required. However DAE systems do not necessarily behave in the same way as ODE systems, as discussed for example by Petzold (1982).

The theory for linear constant-coefficient systems like (3) has long been complete, and is well described by Gantmacher (1959). If the matrix pencil $\{E,A\}$ is singular, then (3) is either underdetermined or inconsistent, so we consider only regular systems, for which $\{E,A\}$ is regular. Then there exist nonsingular matrices P and Q which reduce the pencil to the Kronecker canonical form:

$$P\{E,A\}Q = \left\{ \begin{bmatrix} I_o & & \\ & J_1 & \\ & & J_s \end{bmatrix} , \begin{bmatrix} C & & \\ & I_1 & \\ & & I_s \end{bmatrix} \right\} \tag{36}$$

Where I_o, I_1, ... I_s are unit matrices of orders n_o, n_1, ... n_s, respectively, and J_1, J_2, ... J_s are square matrices with zero elements, except for units on the superdiagonal. The "index of nilpotency" of the pencil is $m = \max_{i=1,2,..s} n_i$.

Using (36) and the transformation:

$$\begin{aligned} [z_o(t), z_1(t), \ldots z_s(t)] &= Q^{-1}x(t) \quad , \\ [g_o(t), g_1(t), \ldots g_s(t)] &= PB\, u(t), \end{aligned} \tag{37}$$

equation (3) yields:

$$\begin{aligned} \dot{z}_o(t) &= C\, z_o(t) + g_o(t) \\ J_i\, \dot{z}_i(t) &= z_i(t) + g_i(t) \ , \ i = 1,2, \ldots s. \end{aligned} \tag{38}$$

Thus the system decomposes into (s+1) independent subsystems. That for $z_o(t)$ is a standard system of ODEs, with a unique solution through any given $z_o(t_o)$. For i=1,2, ... s, we have

$$\begin{aligned} z_i^j(t) &= g_i^j(t) && , \ j=n_i \\ z_i^j(t) &= g_i^j(t) - \dot{z}_i^{j+1}(t) && , \ j=(n_i-1), \ldots 1, \end{aligned} \tag{39}$$

and hence $z_i(t)$ is determined by $g_i(t)$ and its derivatives up to order (n_i-1), with no arbitrary constants. Clearly $z_o(t_o)$ uniquely defines the solution x(t) of equation (3).

For nonlinear DAE systems like (4) it is natural to define a local index at a point in the domain of $f^j(.)$ as the index of the matrix pencil $\{f^j_{\dot{x}^j}, f^j_{x^j}\}$ evaluated at that point. One might then reasonably expect that if equation (4) is satisfied at a certain point, regularity of the pencil at this point would guarantee the existence of a unique solution passing through the point on some neighbourhood of it. Surprizingly this is not the case, but Sargent (1985) proves the following existence theorem:

Theorem

Given a compact set $G \subset R \times R^n \times R^n \times R^m \times R^q \times R^p$ and a function $f(t,\dot{x},x,y,u,v)$: $G \to R^{n+m}$, suppose that $f(\cdot) \varepsilon C^k(G)$ for some $k \geq 1$, and there exists $(t_o,a,b,c,u_o,v_o) \varepsilon \overset{o}{G}$ (the interior of G) such that the matrix $[f_{\dot{x}},f_y]$ is nonsingular at this point and $f(t_o,a,b,c,u_o,v_o)=0$. Then there exist neighbourhoods $T \subset R$ of t_o, $B \subset R^n$ of b, $U \subset R^q$ of u_o and $V \subset R^p$ of v_o, such that for each $u(\cdot) \varepsilon C^{\ell}(T)$ with $u(t_o) \varepsilon U$, each $v \varepsilon V$, and each $x(t_o) \varepsilon B$, there exists a unique solution $(x(\cdot),y(\cdot))$ of the equation

$$f(t,\dot{x}(t),x(t),y(t),u(t),v) = 0, \quad t \varepsilon T, \tag{40}$$

with $(t,\dot{x}(t),x(t),y(t),u(t),v) \varepsilon G$, $t \varepsilon T$, and $(\dot{x}(\cdot),y(\cdot)) \varepsilon C^j(T)$, where $j=\min(k,\ell)$. Moreover if $[f_{\dot{x}},f_y]$ is nonsingular on G, this solution can be extended to the boundary of G.

It also follows easily from the proof that $\dot{x}(\cdot)$, $x(\cdot)$, $y(\cdot)$ are uniformly continuous in $x(t_o)$, v and $u(\cdot)$.

Thus systems satisfying this nonsingularity condition on $[f_{\dot{x}},f_y]$ satisfy all the assumptions underlying the derivations in Section 3. We note that it also guarantees the existence and regularity of solutions of the adjoint system (34). The adjoint boundary conditions (35) will also be well defined if the matrices appearing on the left-hand sides of (35a), (35c) and (35d) are nonsingular, and this provides regularity conditions on the switching functions $h^j(\cdot)$ and the junction conditions (6). In particular, the dimension of $J^j(\cdot)$ must be the same as the subset $x^{j+1}(t)$.

We also note that the appearance of $Df^j(t_j)$ and $Dh^j(t_j)$ in these conditions implies the existence of $\dot{u}^j(t_j)$, and hence that admissible control functions must be differentiable on each subinterval. However, as noted earlier we can accommodate discontinuities in the controls by defining additional stage switching-times to occur at these discontinuities. At such points the subset x(t) is continuous and the junction condition is simply $x^{j+1}(t_j) = x^j(t_j)$; it then follows from (35) that $\mu^j(t_j) = \mu^{j+1}(t_j)$. If the switching-time t_j is an element of v, we also have from equation (28):

$$\phi_{t_j} = \mu^j(t_j)\ \{\dot{x}^j(t_j) - \dot{x}^{j+1}(t_j)\} \tag{41}$$

There is of course no guarantee that there will be feasible choices of $\dot{x}^{j+1}(t_j)$, $x^{j+1}(t_j)$, $u^{j+1}(t_j)$, v for which (4) and (6) are satisfied at the beginning of each stage, but if a solution exists the nonsingularity of the matrix in (35c) will ensure that it is locally unique - and it can be determined numerically using Newton's method. The regularity of the pencil $\{f^j_{\dot{x}^j},\ f^j_{x^j}\}$ similarly ensures that equation (16) has a unique solution at each step for a sufficiently small step-length, and again this solution may be found using Newton's method.

Sargent (1985) shows that nonsingularity of the matrix $[f_{\dot{x}},\ f_y]$, as in the above theorem, not only implies that the matrix pencil $\{[f_{\dot{x}},0],\ [f_x,f_y]\}$ is regular, but also that its index is one and $f_{\dot{x}}$ is of full rank. Thus the theorem covers only a very restricted class of DAE systems. However, although more general DAE systems certainly have solutions, the solutions of their adjoint systems are no longer uniquely defined, and their regularity is thus in doubt. Clearly more work is required in this area.

Nevertheless, it is always possible to reduce higher index systems to systems of index one, and for regularity the corresponding reduced system must satisfy the above theorem. For constant-coefficient systems, the structure of (39) shows that this reduction process must involve differentiation of some of the equations, and indeed the index may be interpreted as the maximum number of successive differentiations required to reduce the system to a set of ordinary differential equations in all the variables. This interpretation of the index carries over directly to nonlinear systems, as illustrated by the following example:

Example: Simple Pendulum

A unit mass is suspended on a string of unit length, yielding the following system of equations:

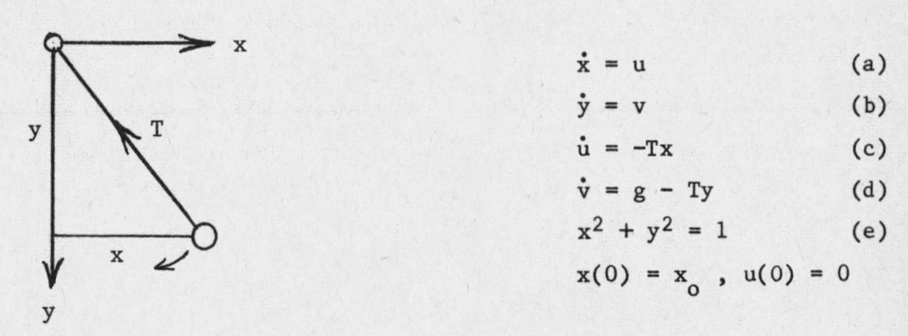

$$\dot{x} = u \tag{a}$$
$$\dot{y} = v \tag{b}$$
$$\dot{u} = -Tx \tag{c}$$
$$\dot{v} = g - Ty \tag{d}$$
$$x^2 + y^2 = 1 \tag{e}$$
$$x(0) = x_o\ ,\ u(0) = 0$$

Of course, the acceleration due to gravity, g, is a given constant, leaving five variables, x, y, u, v, T. We note that only two initial conditions can be specified, although there are four first-order differential equations. The tension T appears only on the right-hand side of these equations, and cannot be determined from the initial conditions, but it is implicitly determined by the algebraic equation (e).

To reduce the system to a set of ODEs, we define new variables: $\dot{u}=p, \dot{v}=q$, (f) then differentiate (e) twice, with appropriate substitutions to yield:

$$xu + y\,v = 0 \qquad (g)$$

$$xp + yq + u^2 + v^2 = 0 \qquad (h)$$

$$-x^2T+y(g-Ty)+u^2+v^2=0 \qquad (i)$$

Differentiation of this last equation would then yield an ODE in $\dot{T}$, so the index is three, but this last differentiation is not necessary for solution of the problem, and in general reduction to an index one system suffices.

For a general nonlinear DAE system, Pantelides (1985) has given an algorithm, using only the occurrence matrix for the system, which determines how many times each equation must be differentiated to reduce the system to one for which $[f_{\dot{x}}, f_y]$ is structurally nonsingular.

A computer program has been written incorporating this algorithm, the necessary algebraic differentiations to effect the reduction, and numerical techniques for solving the resulting optimal control problem, as described in earlier sections.

5. Path Constraints

We should like to extend the algorithm to deal with path constraints of the form:

$$c^j \leq g^j(t,\dot{x}^j(t), x^j(t), u^j(t),v) \leq d^j \;, \quad t \in [t_{j-1},t_j]. \qquad (42)$$

In principle this seems straightforward. The finding of a feasible point at the initial time is simply a matter of solving a nonlinear programme involving (4), (6) and (42). The system equations can then be integrated until one of the constraints in (42) reaches a bound; this is then treated as an equation, implicitly determining one of the controls as integration is continued. The constraint is dropped as an equation if the original specified control would cause it to leave the bound, and Morison (1984) gives an appropriate formula for implementing this test.

This scheme yields a sequence of intervals with a different DAE system in each interval and implicitly determined switching times; it thus has the form of the problem treated in earlier sections. This however requires each DAE system to be reduced to one for which $[f_{\dot{x}}, f_y]$ is structurally nonsingular, and the reduction would have to be carried out, or at least checked, for each interval at each iteration. There is also the problem of associating a control variable with each inequality which will be treated as a state variable when the constraint is active. Fortunately however Pantelides' algorithm is readily extended to select a subset from $\{y(t),u(t)\}$ to be treated as state variables which minimizes the number of differentiations required.

An alternative approach is to convert the inequalities to equations using slack variables. The above reduction of the expanded DAE system then has to be carried out only once, at the beginning. However there can be problems with ill-conditioning if squared slack variables are used to avoid constraints, and otherwise the method relies on a sufficient number of unconstrained control variables. This approach is a natural generalization of the method proposed by Jacobson and Lele (1969) for a single inequality where the "order" of their path constraint is closely related to the index of the corresponding DAE system.

A third approach is to use collocation for integrating the DAE system, as proposed for example by Biegler (1983). Here both state and control variables are parameterized as in (9), and the state equations (4) and path constraints (42) treated as constraints at an appropriate finite set of points. However it will still be necessary to carry out the reduction process for each DAE system along the trajectory to ensure that the number of parameters matches the number of degrees of freedom. There is also the standard problem for collocation methods of choosing the number and positions of points at which constraints are satisfied so that violations elsewhere are within acceptable limits.

Work is continuing on these various approaches, and results will be reported in later publications.

References

1. Biegler, L.T., 1983, "Solution of Dynamic Optimization Problems by Successive Quadratic Programming and Orthogonal Collocation", Comp. and Chem. Eng., 8 (3/4), 243-248.

2. Burrage, K., 1982, "Efficiently Implementable Algebraically Stable Runge-Kutta Methods", SIAM J. Numer. Anal. 19 (2), 245-258, (April, 1982).

3. Cameron, I.T., 1983, "Solution of Differential-Algebraic Systems Using Diagonally Implicit Runge-Kutta Methods", IMA Journal of Numerical Analysis, 3 (3), 273-290, (July, 1983).

4. Cobb, D., 1983, "Descriptor Variable Systems and Optimal State Regulation", IEEE Trans. Auto. Control, AC-28, 601-611.

5. Gantmacher, F.R., 1959, "Applications of the Theory of Matrices", Interscience (New York, 1959).

6. Gear, C.W., 1971, "Simultaneous Numerical Solution of Differential-Algebraic Equations", IEEE Trans. Circuit Theory, CT-18, 89-95.

7. Gear, C.W., and L.R. Petzold, 1984, "ODE Methods for the Solution of Differential-Algebraic Systems", SIAM J. Numer. Anal. 21, 716.

8. Jacobson, D.H., and M.M. Lele, 1969, "A Tranformation Technique for Optimal Control Problems with a State Variable Inequality Constraint", IEEE Trans. Auto. Control, AC-14, 457-464.

9. Pantelides, C.C., 1985, "The Consistent Initialization of Differential-Algebraic Systems", submitted for publication.

10. Petzold, L.R., 1982, "Differential-Algebraic Equations are not ODEs", SIAM J. Sci. Stat. Comput., 3 (3), 367-384, (September 1982).

11. Sargent, R.W.H. and G.R. Sullivan, 1977, "The Development of an Efficient Optimal Control Package", in J. Stoer (ed) "Optimization Techniques - Proceedings of the 8th IFIP Conference on Optimization Techniques, Wurzburg, 1977", Part 2, pp 158-167, Springer-Verlag, (Berlin, 1978).

12. Sargent, R.W.H., 1981, "Recursive Quadratic Programming Algorithms and their Convergence Properties", in J.P. Hennart (ed.), "Numerical Analysis - Proceedings, Cocoyoc, Mexico, 1981", Lecture Notes in Mathematics, pp 208-225, Springer Verlag, (Berlin, 1982).

13. Sargent, R.W.H., 1985, "The Existence and Regularity of Solutions of Differential-Algebraic Systems", submitted for publication.

Polynomial Iteration for Nonsymmetric Indefinite Linear Systems

Howard C. Elman
Yale University
Department of Computer Science
New Haven, CT

Roy L. Streit
Naval Underwater Systems Center
New London, CT

Abstract

We examine iterative methods for solving sparse nonsymmetric indefinite systems of linear equations. Methods considered include a new adaptive method based on polynomials that satisfy an optimality condition in the Chebyshev norm, the conjugate gradient-like method GMRES, and the conjugate gradient method applied to the normal equations. Numerical experiments on several non-self-adjoint indefinite elliptic boundary value problems suggest that none of these methods is dramatically superior to the others. Their performance in solving moderately difficult problems is satisfactory, but for harder problems their convergence is slow.

1. Introduction

In recent years there has been significant progress in the development of iterative methods for solving sparse real linear systems of the form

$$Au = b, \tag{1.1}$$

where A is a nonsymmetric matrix of order N. One key to this progress has been the derivation of polynomial based methods, i.e. methods whose m-th approximate solution iterate has the form

$$u_m = u_0 + q_{m-1}(A)r_0, \tag{1.2}$$

where u_0 is an initial guess for the solution, $r_0 = b - Au_0$, and q_{m-1} is a real polynomial of degree $m-1$. The residual $r_m = b - Au_m$ satisfies

$$r_m = [I - Aq_{m-1}(A)]r_0 = p_m(A)r_0, \tag{1.3}$$

where p_m is a real polynomial of degree m such that $p_m(0) = 1$. Applying any norm to (1.3) gives

$$\|r_m\| \le \|p_m(A)\|\|r_0\|.$$

Moreover, if A is diagonalizable as $A = U\Lambda U^{-1}$, then

$$\|p_m(A)\| = \|Up_m(\Lambda)U^{-1}\| \le \|U\|\|U^{-1}\| \max_{\lambda\in\sigma(A)} |p_m(\lambda)|,$$

The work presented in this paper was supported by the U. S. Office of Naval Research under contract N00014-82-K-0814, by the U. S. Army Research Office under contract DAAG-83-0177 and by the Naval Underwater Systems Center Independent Research Project A70209.

so that

$$\|r_m\| \leq \|U\| \|U^{-1}\| \max_{\lambda \in \sigma(A)} |p_m(\lambda)| \, \|r_0\|. \tag{1.4}$$

Thus any polynomial p_m that is sufficiently small on the eigenvalues of A is a good candidate for generating an iterative method.

The conjugate gradient and Chebyshev methods are well-known polynomial-based methods for solving symmetric positive-definite systems for which the residual polynomials $\{p_m\}$ have desirable optimality properties [8]. Generalizations of these techniques have been developed for solving both symmetric indefinite systems (see e.g. [3, 4, 17, 18]), and nonsymmetric systems with definite symmetric part $(A + A^T)/2$ (see e.g. [5, 8, 14] and references therein). In the latter case, all of the eigenvalues of A lie in either the right half or the left half of the complex plane. Sparse linear systems that both are nonsymmetric and have indefinite symmetric part arise in numerous settings. Examples include the discretization of the Helmholtz equations for modelling acoustic phenomena [1] and the discretization of the coupled partial differential equations arising in numerical semiconductor device simulation [12]. Gradient methods that have been proposed as solvers for such problems include the conjugate gradient method applied to the normal equations (CGN) [9], the biconjugate gradient method [7], the restarted generalized minimum residual method (GMRES) [20], and new methods presented in [11, 26]. Smolarski and Saylor [22] and Saad [19] have proposed adaptive polynomial iteration methods of the form (1.2) using polynomials that are optimal with respect a weighted least squares norm. In this paper, we introduce a polynomial-based method, PSUP, that computes a polynomial that is nearly optimal with respect to the Chebyshev norm on a region containing the eigenvalue estimates and then uses this polynomial in (1.2). We compare its performance with the two gradient methods CGN and GMRES.

In Section 2, we give a brief description of the gradient methods CGN and GMRES. In Section 3, we describe the new PSUP method and several heuristics developed to improve its performance. In Section 4, we describe numerical experiments in which these three methods are used to solve some non-self-adjoint indefinite elliptic problems, and in Section 5 we draw conclusions based on the numerical tests.

2. Gradient Methods

In this section we briefly review two conjugate gradient-like methods for solving nonsymmetric indefinite systems. The conjugate gradient method [9] is applicable only to symmetric positive definite linear systems. For nonsymmetric systems, it can be used to solve the normal equations $A^T A x = A^T b$. The scaled residuals $\{A^T r_m\}$ satisfy

$$A^T r_m = p_m(A^T A) A^T r_0,$$

where p_m is the unique polynomial of degree m such that $p_m(0) = 1$ and $\|r_m\|_2$ is minimum. As is well known, the condition number of $A^T A$ is the square of that of A. Moreover, the standard implementation of CGN requires two matrix-vector products at each iteration, one by A and one by A^T, plus 5N additional operations. The storage requirement is 4N words. The dependence of CGN on $A^T A$ has led to efforts to find alternatives that are more rapidly convergent and less expensive per step. For nonsymmetric systems with positive definite symmetric part, several methods have been shown to be superior to CGN [5].

GMRES is a method proposed for solving nonsymmetric indefinite systems that avoids the use of the normal equations [20]. Given an initial guess, u_0, for the solution, with residual r_0, this method generates an orthogonal basis $\{v_1, \ldots, v_m\}$ for the Krylov space

$$K_m = \text{span}\{r_0, Ar_0, \ldots, A^{m-1} r_0\}$$

using Arnoldi's method. Let $v_1 = r_0/\|r_0\|_2$. The Arnoldi process computes for $j = 1, \ldots, m$

$$h_{ij} = (Av_j, v_i), \qquad i = 1, \ldots, j,$$

$$\hat{v}_{j+1} = Av_j - \sum_{i=1}^{j} h_{ij} v_i,$$

$$h_{j+1,j} = \|\hat{v}_{j+1}\|_2,$$

$$v_{j+1} = \hat{v}_{j+1}/h_{j+1,i}.$$

GMRES then computes an approximate solution

$$u_m = u_0 + \sum_{j=1}^{m} \alpha_j v_j, \tag{2.1}$$

where the scalars $\{\alpha_j\}_{j=1}^m$ are chosen so that $\|r_m\|_2$ is minimum. These scalars can be computed by solving the upper Hessenberg least squares problem

$$\min_{\alpha} \left\| \|r_0\|_2 e_1 - \hat{H}_m \alpha \right\|_2,$$

where $e_1 = (1, 0, \ldots, 0)^T \in \mathbf{R}^{m+1}$ and $\hat{H}_m$ is the Hessenberg matrix of size $(m+1) \times m$ whose (i,j)-entry is h_{ij} [20]. By the choice of basis and the minimization property, $r_m = p_m(A) r_0$ where p_m is the real polynomial of degree m such that $p_m(0) = 1$ and p_m is optimal with respect to the residual norm $\|r_m\|_2$ (c.f. [8] for other formulations of this optimal iteration).

In a practical implementation, the dimension m of the Krylov space is fixed, and the GMRES iteration is restarted with u_m in place of u_0. This is the GMRES(m) method. Defining one "step" to be the average of the m-fold iteration divided by m, the cost per step is $(m + 3 + 1/m)N$ operations plus one matrix-vector product. It requires $(m+2)N$ words of storage.

We remark that the Arnoldi process was originally developed as a technique for computing eigenvalues [27]. Let V_m denote the matrix whose columns are the m vectors generated by the Arnoldi step in GMRES(m), and let H_m denote the square upper Hessenberg matrix consisting of the first m rows of $\hat{H}_m$. Then V_m is an orthonormal matrix of order $N \times m$ that satisfies

$$V_m^T A V_m = H_m. \tag{2.2}$$

Relation (2.2) resembles a similarity transformation, and Arnoldi's method consists of using the eigenvalues of H_m as estimates for (some of) the eigenvalues of A. Suppose $A = U \Lambda U^{-1}$ for diagonal Λ and r_0 is dominated by m eigenvectors $\{u_j\}_{j=1}^m$, with corresponding eigenvalues $\{\lambda_j\}_{j=1}^m$. Then the residual after m GMRES steps satisfies [6]

$$\|r_m\|_2 \le \|U\|_2 \|U^{-1}\|_2 \; c_m \; \|e\|_2$$

where

$$c_m = \max_{k>m} \prod_{j=1}^{m} |\lambda_k - \lambda_j| / |\lambda_j|$$

and e is orthogonal to $\{u_j\}_{j=1}^m$. Loosely speaking, GMRES(m) damps out from the residual the eigenvectors whose eigenvalues are computed by Arnoldi's method.

3. The PSUP Method

The gradient methods just described compute iterates and residuals that satisfy (1.2) and (1.3) (for CGN, with respect to $A^T A$) in which the polynomials are built up recursively without explicit computation of their coefficients. In this section, we describe an alternative iteration that computes explicitly the coefficients of a polynomial $q_{m-1}(z)$ for which $p_m(z) = 1 - zq_{m-1}(z)$ is small on the spectrum $\sigma(A)$. In the following, we will refer to the polynomial $q_{m-1}(z)$ of (1.2) as the "iteration polynomial" and to the polynomial $p_m(z) = 1 - zq_{m-1}(z)$ of (1.3) as the "residual polynomial."

Suppose a compact region $D \subset \mathbf{C}$ contains $\sigma(A)$. Let p_m be a polynomial of degree m that satisfies

$$p_m(0) = 1, \qquad \|p_m\| = \max_{z \in D} |p_m(z)| = \epsilon < 1.$$

As is evident from (1.4), an iteration having p_m as its residual polynomial will result in a decrease of the residual norm if ϵ is small enough. The best possible iteration polynomial with respect to this norm (the Chebyshev norm) is the solution to the minimax problem

$$\epsilon = \min_{q_{m-1}} \max_{z \in D} |1 - zq_{m-1}(z)|. \tag{3.1}$$

Let $q_{m-1}(z) = \sum_{j=0}^{m-1} a_j z^j$. The solution to (3.1) is also the Chebyshev solution to the infinite system of equations

$$\sum_{j=0}^{m-1} z^{j+1} a_j = 1, \qquad z \in \partial D. \tag{3.2}$$

Only the boundary ∂D need be considered because of the maximum modulus principle.

The PSUP method uses an iteration polynomial obtained from an approximate solution to (3.1). We briefly summarize the technique used; details can be found in [24]. First, (3.2) is replaced by a finite dimensional problem

$$\sum_{j=0}^{m-1} z^{j+1} a_j = 1, \qquad z \in \partial D_M, \tag{3.3}$$

where ∂D_M is a finite subset of ∂D containing M points, $M > m$. Equation (3.3) is an overdetermined system of M equations in the m unknowns $\{a_j\}_{j=0}^{m-1}$. The Chebyshev problem for (3.3) is given by

$$\min_{\{a_j\}} \max_{z \in \partial D_M} \left| \sum_{j=0}^{m-1} z^{j+1} a_j - 1 \right|. \tag{3.4}$$

Second, equation (3.4) is solved approximately using a semi-infinite linear programming approach to complex approximation, which is based on the identity $|w| = \max_{0 \leq \theta < 2\pi} Re(we^{-i\theta})$, $w \in \mathbf{C}$. Let $\Theta = \{\theta_1, \ldots, \theta_p\} \subset [0, 2\pi)$, and define the *discretized absolute value*

$$|w|_\Theta = \max_{\theta \in \Theta} \; Re(we^{-i\theta}).$$

Consider the discretized problem

$$\min_{\{a_j\}} \max_{z \in \partial D_M} \left| \sum_{j=0}^{m-1} z^{j+1} a_j - 1 \right|_\Theta, \tag{3.5}$$

where the absolute value in (3.4) is replaced by the discretized absolute value. This gives rise to a linear program for $\{a_j\}_{j=0}^{m-1}$. Let ϵ^* denote the minimax value of $|\sum_{j=0}^{m-1} z^{j+1} a_j - 1|$ at the solution to (3.4), and let ϵ_p^* denote the minimax value for (3.5). It can be shown that

$$|w|_\Theta \leq |w| \leq |w|_\Theta \; sec(\alpha/2)$$

for all $w \in \mathbf{C}$, and consequently that

$$\epsilon_p^* \leq \epsilon^* \leq \epsilon_p^* \, sec(\alpha/2),$$

where α is the smallest difference (mod 2π) between two neighboring angles in Θ. The upper bounds are sharpest for given p when Θ consists of the p-th roots of unity, so that $\alpha = 2\pi/p$. We use this choice of Θ in the following, with $p = 256$ so that $sec(\alpha/2) = 1.000075$.

The dual of the LP (3.5) can be written in the form

$$\min_{S\in\mathbf{R}^{M\times p},\ Q\in\mathbf{R}} \quad Re[e_M^T S e^{-i\Theta}]$$

$$\text{subject to: } S \geq 0, \quad Q \geq 0, \quad Z^T S e^{-i\Theta} = 0 \in \mathbf{C}^m$$

$$\text{and} \quad Q + \sum_{j=1}^{M} \sum_{k=1}^{p} S_{jk} = 1,$$

where $e_M \in \mathbf{C}^M$ is the vector whose components are all 1, $Z \in \mathbf{C}^{M\times m}$ is the coefficient matrix of (3.4), and $e^{-i\Theta} \in \mathbf{C}^p$ denotes the vector whose j-th component is $e^{-i\theta_j}$. Q is a slack variable which must be 0 if $\epsilon_p^* > 0$. A straightforward application of the simplex method to the dual requires $O(Mmp)$ multiplications per simplex iteration and $O(Mmp)$ storage locations. In [24], it is shown that the factor p can be eliminated from these estimates by exploiting the special structure of the dual. These economies leave unaltered the sequence of basic feasible solutions that the simplex method generates en route to the solution. Moreover, they simplify further if the coefficients $\{a_j\}$ are required to be real. In practice the number of simplex iterations has been observed to be $O(m)$ so that the computational effort to compute $\{a_j\}$ using the algorithm in [23] is $O(Mm^2)$. In the experiments discussed below, both M and m are significantly smaller than the order N of the linear system so that construction of the coefficients of the iteration polynomial is a low order cost of the solution process.

Given u_0 and r_0, the basic PSUP iteration consists of repeated application of the iteration polynomial q_{m-1}, as follows:

Algorithm 1: The PSUP iteration.
For $k = 1, 2, \ldots$ Do
$u_{km} = u_{(k-1)m} + q_{m-1}(A) r_{(k-1)m}$
$r_{km} = b - A u_{(k-1)m}$.

The actual computation $w \leftarrow q_{m-1}(A)r$ is performed using Horner's rule:

$w \leftarrow a_{m-1} r$
For $j = 1$ to $m-1$ Do
$v \leftarrow Aw$
$w \leftarrow a_{m-1-j} r + v$.

The m-fold PSUP iteration requires m matrix-vector products and m scalar vector products, so that the "average" cost is one matrix-vector product and one scalar-vector product. PSUP requires $4N$ storage, for u, r, v and w.

In practice, the PSUP iteration needs estimates of the eigenvalues of A in order to obtain the set D. Several adaptive techniques have been developed for combining an eigenvalue estimation procedure with polynomial iteration [6, 13, 19]. We will use the hybrid technique developed in [6, 19], which uses Arnoldi's method for eigenvalue estimates.

First, the Arnoldi process is used to compute some number k_i of eigenvalue estimates prior to execution of the PSUP iteration. Given these estimates, a set D is constructed that contains them, from which the PSUP iteration polynomial q_{m-1} is computed. (We discuss our choice for D below.) One possible strategy is to perform the PSUP iteration with q_{m-1}

until the iteration converges. However, there is no guarantee that all the extreme eigenvalues of A are computed by the Arnoldi procedure. The set D is contained in the lemniscate region [10] $L_m = \{z \in \mathbf{C} \mid |p_m(z)| \le \epsilon\}$, where ϵ and $p_m = 1 - zq_{m-1}(z)$ solve (3.1). Moreover, the modulus of p_m is greater than ϵ outside L_m and tends to grow rapidly outside L_m, at least in some directions. If an eigenvalue λ lies outside L_m and $|p_m(\lambda)|$ is large enough, then the PSUP method will diverge.

One way to avoid this behavior is to invoke the adaptive procedure: if PSUP diverges then k_a additional Arnoldi steps are performed to compute k_a new eigenvalue estimates. These estimates are then used to construct a new enclosing set D and a new iteration polynomial q_{m-1}, with which the PSUP iteration is resumed. A good choice for a starting vector v_1 is the last residual from the previous PSUP iteration (normalized to have unit norm). For if PSUP diverges, then the residual will tend to be dominated by the eigenvectors whose eigenvalues are not being damped out by the PSUP polynomial. Moreover, this technique can be improved using GMRES. Once the k_a Arnoldi vectors are available, the GMRES(k_a) iteration (2.1) can be performed at relatively little extra expense. This has the effect of damping out from the residual the eigenvector components that were being enhanced by the previous PSUP iteration.

Rather than use the PSUP iteration alone, we consider a hybrid PSUP-GMRES method that makes use of these observations. This method consists of repeated iteration of some number s of PSUP steps, followed by a smaller number k_a of Arnoldi-GMRES steps. The initial eigenvalue estimates are provided by k_i Arnoldi-GMRES steps, where k_i may differ from k_a. In addition, the adaptive procedure is invoked immediately if the residual norm of the PSUP iteration increases by some tolerance τ relative to the smallest residual previously encountered. The following is a modification of the hybrid method developed in [6] that uses the PSUP iteration:

Algorithm 2: The hybrid GMRES-PSUP method.
Choose u_0. Compute $r_0 = b - Au_0$.
Until Convergence Do
Adaptive (Initialization) Steps: Set $v_1 =$ the current normalized residual,
perform k_a (or k_i) Arnoldi/GMRES steps, and use the new eigenvalue
estimates to update (or initialize) the PSUP coefficients.
PSUP Steps: While $(\|r_j\|/\|r_{min}\| \le \tau))$
Perform s steps of the PSUP iteration (Algorithm 1) to
update the approximate solution u_j and residual r_j.

For the enclosing set D we take the union of the four sets D_j, where D_j is the convex hull of the set of eigenvalue estimates in the j-th quadrant of the complex plane. With this choice, if the extreme eigenvalues of each quadrant have been computed, then all the eigenvalues are contained in D. If all the eigenvalue estimates in either half plane are real, then the part of D containing these estimates is taken to be the line segment between the leftmost and rightmost estimates in the half plane.

There is no guarantee that the eigenvalue estimates computed by Arnoldi's method are accurate. Moreover, since the PSUP residual polynomial has the value 1 at the origin, if D contains points with both positive and negative real parts that are near the origin, then the Chebyshev norm of the residual polynomial will be very close to 1. (See Section 4 for an example.) We consider one heuristic designed to improve the performance of the hybrid PSUP method on problems with eigenvalues very near the origin: we successively remove the points closest to the origin from the set of eigenvalue estimates (and generate a smaller D) until the norm of the PSUP polynomial is smaller than some predetermined value η, and use that polynomial for the PSUP iteration.

There are two possible effects of this heuristic. If the deleted points are not accurate as eigenvalue estimates, then the resulting PSUP iteration will be just as robust and more rapidly convergent than if the deleted points had been included. On the other hand, if the deleted points are good estimates, then the PSUP polynomial will probably be large on the deleted points, and the iteration will not damp out the residual in the direction of the corresponding eigenvectors. However, if the dimension of this eigenspace is small (say, 2 or 3), then the iteration should damp out the residual in all other components, so that the residual should be dominated by a small number of components. In this situation, a small number of GMRES steps should damp out these dominant components. We will refer to the hybrid PSUP method with this heuristic added as the GMRES/Reduced-PSUP scheme.

We note that with the methods of [24], (3.5) can be also solved with the constraint

$$\max_{z\in E}\left|\sum_{j=0}^{m-1} z^{j+1}a_j - 1\right|_\Theta \le 1,$$

where E is some finite set. In particular, if E is the set of deleted eigenvalue estimates in the GMRES/Reduced-PSUP scheme, then the PSUP polynomial on the reduced set D can be forced to be bounded in modulus by one on the deleted points. In experiments with this version of the GMRES/Reduced-PSUP iteration, we found its performance to be essentially the same as that of the unconstrained version described above.

4. Numerical Experiments

In this section, we compare the performance of CGN, GMRES(m), GMRES/PSUP and GMRES/Reduced-PSUP in solving several linear systems arising from a finite difference discretization of the differential equation

$$-\Delta u + 2P_1u_x + 2P_2u_y - P_3u = f, \quad u \in \Omega, \tag{4.1}$$

$$u = g, \quad u \in \partial\Omega,$$

where Ω is the unit square $\{0 \le x, y \le 1\}$, and P_1, P_2 and P_3 are positive parameters. We use $f = g \equiv 0$, so that the solution to (4.1) is $u = 0$.

We discretize (4.1) by finite differences on a uniform $n \times n$ grid, using centered differences for the Laplacian and the first derivatives. Let $h = 1/(n+1)$. After scaling by h^2, the matrix equation has the form (1.1) in which the typical equation for the unknown $u_{ij} \approx u(ih, jh)$ is

$$(4-\sigma)u_{ij} - (1+\beta)u_{i-1,j} + (-1+\beta)u_{i+1,j} - (1+\gamma)u_{i,j-1} + (-1+\gamma)u_{i,j+1} = h^2 f_{ij},$$

where $\beta = P_1h$, $\gamma = P_2h$, $\sigma = P_3h^2$ and $f_{ij} = f(ih, jh)$. The eigenvalues of A are given by [21]

$$4-\sigma + 2\sqrt{1-\beta^2}\, cos\frac{s\pi}{n+1} + 2\sqrt{1-\gamma^2}\, cos\frac{t\pi}{n+1}, \qquad 1 \le s,t \le n.$$

The eigenvalues of the symmetric part are

$$4-\sigma + 2cos\frac{s\pi}{n+1} + 2cos\frac{t\pi}{n+1}, \qquad 1 \le s,t \le n.$$

The leftmost eigenvalue of the symmetric part, corresponding to $s = t = n$, is given by

$$(2\pi^2 - P_3)h^2 + O(h^4),$$

so that for small enough h the symmetric part is indefinite when $P_3 > 2\pi^2$.

Six test problems corresponding to six choices of the parameter set $\{P_1,\ P_2,\ P_3\}$ are considered. We use the three values $P_3 = 30$, 80, and 250 together with each of the pairs of values $\{P_1 = 1,\ P_2 = 2\}$ and $\{P_1 = 25,\ P_2 = 50\}$. For all tests, $n = 31$, so that the order $N = n^2$ is 969. For all six test problems, the coefficient matrix A is indefinite, and the number of negative eigenvalues of $(A + A^T)/2$ is increasing as P_3 grows. For the first choice of the (P_1, P_2) pair, A is mildly nonsymmetric and its eigenvalues are real, and for the second choice, A is more highly nonsymmetric and has complex eigenvalues.

Although it is not our intention here to examine preconditioners for indefinite systems, preconditioning has been shown to be a critical factor in the performance of iterative methods [3, 5, 15]. In our tests, we precondition (1.1) by the finite difference discretization of the Laplacian. That is, the iterative methods being considered are applied to the *preconditioned problem*

$$AQ^{-1}\hat{x} = b, \quad x = Q^{-1}\hat{x},$$

where Q is the discrete Laplacian. (See [2] for an asymptotic analysis of this preconditioner for finite element discretizations.) The preconditioned matrix-vector product then consists of a preconditioning solve of the form $Q^{-1}v$ and a matrix multiply of the form Av. Since Ω is a square domain, the preconditioning is implemented using the block cyclic reduction method at a cost of $3n^2 \log_2 n$ operations [25]. We have confirmed numerically that the preconditioned matrix AQ^{-1} in all six problems has indefinite symmetric part.

We use the following parameters for the hybrid GMRES-PSUP iteration. In an effort to obtain the dominant and subdominant eigenvalues of each quadrant at the outset, the initialization step consists of eight GMRES steps ($k_i = 8$) giving eight eigenvalue estimates. All subsequent calls to the adaptive procedure consist of four GMRES steps ($k_a = 4$). For all tests with PSUP, we use a residual polynomial of degree four ($m = 4$), and allow at most $s = 32$ PSUP steps (or eight successive applications of the PSUP polynomial). The adaptive procedure is invoked if the residual norm increases during a PSUP step ($\tau = 1$), or after s steps are performed. We use $M = 100$ points for the discretized enclosing set ∂D_M, and allocate them so that the number of points in each quadrant is approximately proportional to the circumference of the convex hull in that quadrant. For subsets of D that overlap on quadrant boundaries (e.g. if a line segment on the real line is shared by regions in the first and fourth quadrants), the shared boundary is discretized twice. For the GMRES/Reduced-PSUP scheme, in which eigenvalue estimates closest to the origin are deleted until the minimax norm is less than some tolerance η, we examine $\eta = .5$ and $.3$. For this scheme, we take k_a to be two plus the number of eigenvalue estimates deleted. We use the notation GMRES-PSUP(m) (with $m = 4$) for the "unreduced" scheme, and GMRES-PSUP(m, η) for the reduced version.

We examine GMRES(m) for $m = 5$ and $m = 20$. Recall that the latter version generates a higher degree optimal polynomial at the expense of a larger average cost per step.

All numerical tests were run on a VAX 11-780 in double precision (55 bit mantissa). The initial guess in all runs was a vector u_0 of random numbers between -1 and 1. Figures 1 - 6 show the performance of the methods measured in terms of multiplication counts, for the six problems (also numbered 1 - 6). Note that the horizontal scale of Figure 1 is wider than the others, and the scales in Figures 5 and 6 are slightly narrower. Table 1 shows the iteration counts needed to satisfy the stopping criterion of

$$\frac{\|r_j\|_2}{\|r_0\|_2} \le 10^{-6}.$$

A maximum of 100, 150, and 200 iterations were permitted for the CGN, GMRES and PSUP methods, respectively. (For these iteration counts, CGN, GMRES(20) and GMRES-PSUP(4) performed roughly the same number of operations.) Our main observations on this data are:

1. Problems 1 and 3 are solved efficiently by nearly all the methods, but for the other four problems convergence is slow.

2. In general, the hybrid GMRES-PSUP(m) scheme is weakest. The plateaus in Figures 3, 5 and 6 for this method correspond to the PSUP step, for which convergence is very slow. The "reduction" heuristic improves the performance, but the improvement is due largely to increased effectiveness of the GMRES part of the iteration (e.g. in the steep drops of Figures 2 - 4), and the improved performance is not better than that of GMRES alone.
3. On the whole, GMRES(20) and CGN are the most effective methods for these problems, but they are not dramatically superior to the others. GMRES(20) converges more rapidly than GMRES(5).

Excluding storage for the matrix and right hand side, the storage requirements for the methods considered are

CGN:	$4N$
GMRES(5):	$7N$
GMRES(20):	$22N$
All PSUP variants:	$10N$

The high cost of the PSUP methods is due to the eight initializing GMRES steps.

Although the GMRES/Reduced-PSUP (PSUP(m,η)) scheme is not as fast as pure GMRES, the reduction heuristic does have its intended effect of improving upon the hybrid scheme. We briefly examine the effect of the heuristic on Problem 3, focusing on two curve segments of Figure 3: the plateau of curve D (GMRES-PSUP(4)) between multiplication counts 200000 and 300000, and the last plateau in curve E (GMRES-PSUP(4,.5)). For curve D, on return from the adaptive step at about multiplication count 200000, the real parts of the eigenvalue estimates lie in the intervals [-3,-.33] and [0.4,.98], the Chebyshev norm of the residual polynomial is .98, and convergence is slow. For curve E, on return from the adaptive step prior to the last plateau of the curve, the real parts of the eigenvalue estimates lie in the intervals [-3,-.56] and [.05,.97], and the Chebyshev norm is .96. The effect of deletion of points is shown in Table 2. The Chebyshev norm is very large when there are points near the origin, and it declines as these points are deleted. The deletion of points does not significantly hurt the PSUP part of the iteration and it strongly enhances the effect of the GMRES steps.

Problem #	1	2	3	4	5	6
CGN	13	>100	28	>100	>100	>100
GMRES(5)	13	>150	46	>150	>150	>150
GMRES(20)	10	111	17	119	>150	>150
GMRES-PSUP	16	>200	199	>200	>200	>200
PSUP(4,.5)	16	>200	62	>200	>200	>200
PSUP(4,.3)	16	>200	70	>200	>200	>200

Table 1: Iteration counts.

Deleted Points	Intervals Containing Real Parts	Chebyshev Norm
-	[-3, -.56], [.05,.97]	.96
.05	[-3, -.56], [.34,.97]	.76
.34	[-3, -.56], [.61,.97]	.55
-.56	[-3,-1.46], [.61,.97]	.33

Table 2: Effect of point deletion on GMRES/Reduced-PSUP(4,.5) for Problem 3.

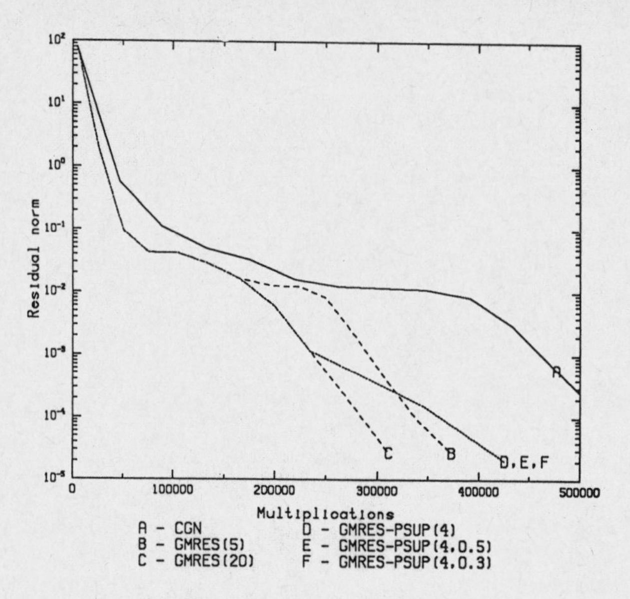

Figure 1: $P_1 = 1$, $P_2 = 2$, $P_3 = 30$

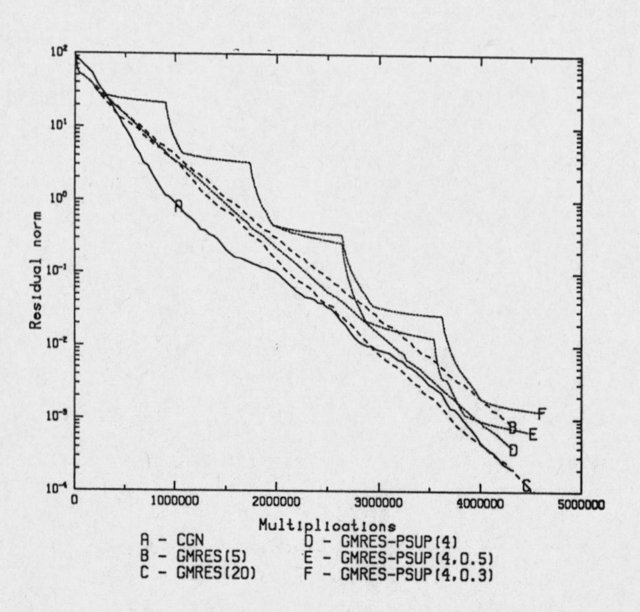

Figure 2: $P_1 = 25$, $P_2 = 50$, $P_3 = 30$

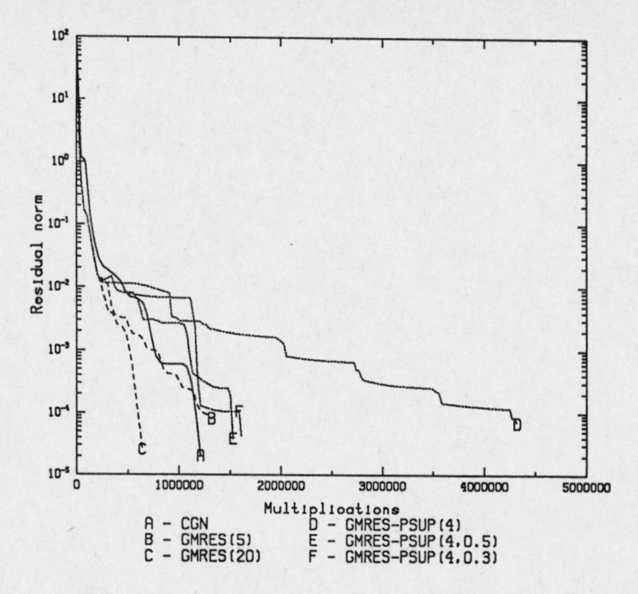

Figure 3: $P_1 = 1$, $P_2 = 2$, $P_3 = 80$

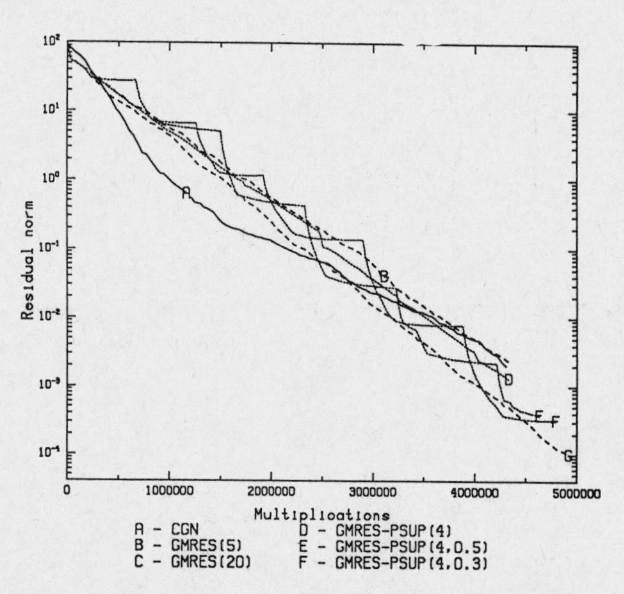

Figure 4: $P_1 = 25$, $P_2 = 50$, $P_3 = 80$

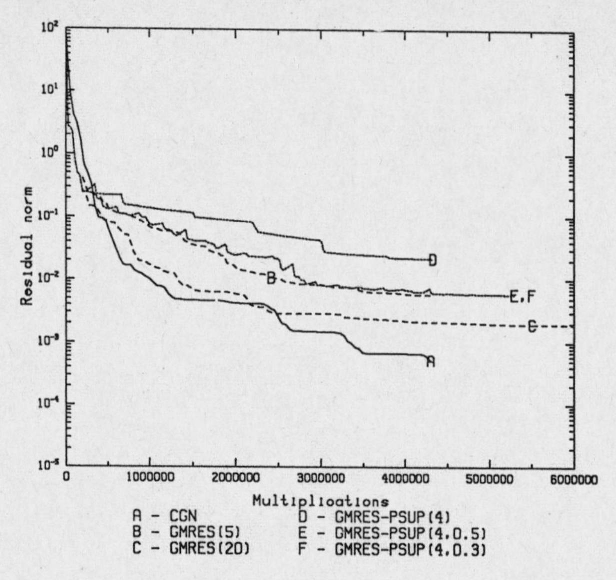

Figure 5: $P_1 = 1,\ P_2 = 2,\ P_3 = 250$

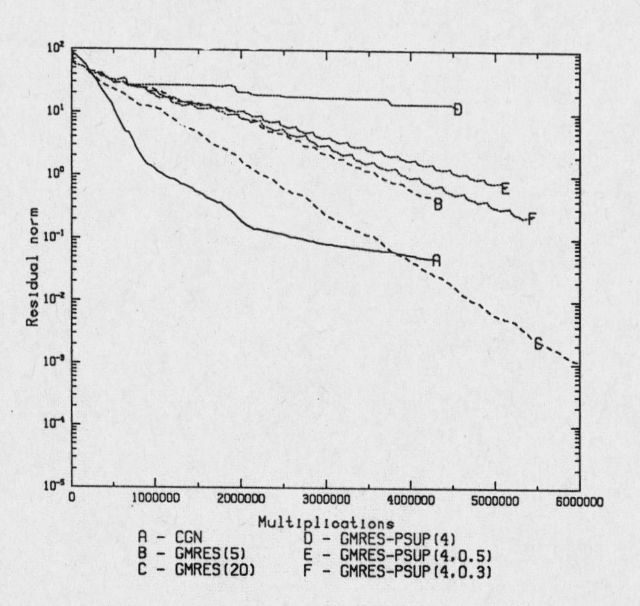

Figure 6: $P_1 = 25,\ P_2 = 50,\ P_3 = 250$

We remark that we also considered other variants of the PSUP iteration. In experiments with degrees $m = 6$ and 10 the performance of PSUP was essentially the same.* Moreover, as we noted in Section 3, a variant of the GMRES/Reduced-PSUP in which the PSUP polynomial is constrained to be bounded in modulus by one on the set of deleted eigenvalue estimates displayed about the same behavior as the unconstrained version. Similarly, we tested LSQR [16], a stabilized version of CGN, and found that its performance was nearly identical to CGN.

5. Conclusions

The GMRES and PSUP methods are iterative methods that are optimal in the class of polynomial-based methods with respect to the Euclidean or l_∞ norms respectively, for arbitrary nonsingular linear systems. For linear systems in which the coefficient matrix is either symmetric or definite (or both), these types of methods are effective solution techniques [3, 5]. In particular, they are superior to solving the normal equations by the conjugate gradient method. In the results of Section 4, the methods based on polynomials in the coefficient matrix are not dramatically superior to CGN, especially for systems that are both highly nonsymmetric and highly indefinite. GMRES appears to be a more effective method than PSUP.

We note that the best results for other classes of problems depend strongly on preconditioning. We used the discrete Laplacian as a preconditioner in our experiments, and the large iteration/work counts in the results show that this is not a good choice for the given mesh size when the coefficients in the differential operator are large. We believe that improvements in preconditioners are needed to handle this class of problems.

---

*In some tests with degree 16, we were unable to generate the polynomial coefficients. We believe the choice of the powers of z as basis functions makes (3.5) ill conditioned for large m; see [19]. In addition, the implementation based on Horner's rule may suffer from instability for large m.

References

[1] A. Bayliss, C. I. Goldstein and E. Turkel, *An iterative method for the Helmholtz equation,* Journal of Computational Physics, 49 (1983), pp. 443–457.

[2] J. H. Bramble and J. E. Pasciak, Preconditioned iterative methods for nonselfadjoint or indefinite elliptic boundary value problems, H. Kardestuncer ed., *Unification of Finite Element Methods,* Elsevier Science Publishers, New York, 1984, pp. 167–184.

[3] R. Chandra, *Conjugate Gradient Methods for Partial Differential Equations,* Ph.D. Thesis, Department of Computer Science, Yale University, 1978. Also available as Technical Report 129.

[4] C. de Boor and J. R. Rice, *Extremal polynomials with application to Richardson iteration for indefinite linear systems,* SIAM J. Sci. Stat. Comput., 3 (1982), pp. 47–57.

[5] H. C. Elman, *Iterative Methods for Large, Sparse, Nonsymmetric Systems of Linear Equations,* Ph.D. Thesis, Department of Computer Science, Yale University, 1982. Also available as Technical Report 229.

[6] H. C. Elman, Y. Saad and P. E. Saylor, *A Hybrid Chebyshev Krylov-Subspace Method for Nonsymmetric Systems of Linear Equations,* Technical Report YALEU/DCS/TR-301, Yale University Department of Computer Science, 1984. To appear in SIAM J. Sci. Stat. Comput.

[7] R. Fletcher, Conjugate gradient methods for indefinite systems, G. A. Watson ed., *Numerical Analysis Dundee 1975,* Springer-Verlag, New York, 1976, pp. 73–89.

[8] L. A. Hageman and D. M. Young, *Applied Iterative Methods,* Academic Press, New York, 1981.

[9] M. R. Hestenes and E. Stiefel, *Methods of conjugate gradients for solving linear systems,* Journal of Research of the National Bureau of Standards, 49 (1952), pp. 409–435.

[10] E. Hille, Volume II: *Analytic Function Theory,* Blaisdell, New York, 1962.

[11] K. Ito, *An Iterative Method for Indefinite Systems of Linear Equations,* Technical Report NAS1-17070, ICASE, April 1984.

[12] T. Kerkhoven, *On the Choice of Coordinates for Semiconductor Simulation,* Technical Report RR-350, Yale University Department of Computer Science, 1984.

[13] T. A. Manteuffel, *Adaptive procedure for estimation of parameters for the nonsymmetric Tchebychev iteration,* Numer. Math., 31 (1978), pp. 187–208.

[14] ———, *The Tchebychev iteration for nonsymmetric linear systems,* Numer. Math., 28 (1977), pp. 307–327.

[15] J. A. Meijerink and H. A. van der Vorst, *An iterative solution method for linear systems of which the coefficient matrix is a symmetric M-matrix,* Math. Comp., 31 (1977), pp. 148–162.

[16] C. C. Paige and M. A. Sanders, *LSQR: An algorithm for sparse linear equations and sparse least squares,* ACM Trans. on Math. Software, 8 (1982), pp. 43–71.

[17] C. C. Paige and M. A. Saunders, *Solution of sparse indefinite systems of linear equations,* SIAM J. Numer. Anal., 12 (1975), pp. 617–629.

[18] Y. Saad, *Iterative solution of indefinite symmetric systems by methods using orthogonal polynomials over two disjoint intervals,* SIAM J. Numer. Anal., 20 (1983), pp. 784–811.

[19] ———, *Least squares polynomials in the complex plane with applications to solving sparse nonsymmetric matric problems,* Technical Report 276, Yale University Department of Computer Science, June 1983.

[20] Y. Saad and M. H. Schultz, *GMRES: A Generalized Minimal Residual Algorithm for Solving Nonsymmetric Linear Systems,* Technical Report 254, Yale University Department of Computer Science, 1983.

[21] G. D. Smith, *Numerical Solution of Partial Differential Equations: Finite Difference Methods,* Oxford University Press, New York, 1978.

[22] D. C. Smolarski and P. E. Saylor, *Optimum Parameters for the Solution of Linear Equations by Richardson's Iteration,* May 1982. Unpublished manuscript.

[23] R. L. Streit, *An Algorithm for the Solution of Systems of Complex Linear Equations in the l_∞ Norm with Constraints on the Unknowns,* 1983. Submitted to ACM Trans. on Math. Software.

[24] ———, *Solution of Systems of Complex Linear Equations in the l_∞ Norm with Constraints on the Unknowns,* Technical Report 83-3, Systems Optimization Laboratory, Stanford University Department of Operations Research, 1983. To appear in SIAM J. Sci. Stat. Comput.

[25] P. N. Swarztrauber, *The methods of cyclic reduction, Fourier analysis and the FACR algorithm for the discrete solution of Poisson's equation on a rectangle,* SIAM Review, 19 (1977), pp. 490–501.

[26] M. A. Saunders, H. D. Simon, and E. L. Yip, *Two Conjugate-Gradient-Type Methods for Sparse Unsymmetric Linear Equations,* Technical Report ETA-TR-18, Boeing Computer Services, June 1984.

[27] J. H. Wilkinson, *The Algebraic Eigenvalue Problem,* Oxford University Press, London, 1965.

VIEWING THE CONJUGATE GRADIENT METHOD AS A TRUST REGION ALGORITHM*

Jorge Nocedal
Department of Electrical Engineering
and Computer Science
Northwestern University
Evanston, IL 60201

We are interested in solving the unconstrained optimization problem

$$\min f(x) \ , \tag{1}$$

where $f: \mathbb{R}^n \to \mathbb{R}$ is twice continuously differentiable. The conjugate gradient method for solving (1) is given by

$$d_K = -g_K + \frac{g_K^T y_{K-1}}{y_{K-1}^T d_{K-1}} d_{K-1} \tag{2a}$$

$$x_{K+1} = x_K + \alpha_K d_K, \tag{2b}$$

where α_K is a steplength, $g_K = \nabla f(x_K)$ and $y_{K-1} = g_K - g_{K-1}$. The algorithm is started by choosing an initial point x_0 and setting $d_0 = -g_0$.

The conjugate gradient method is useful in solving large problems because it does not require matrix storage. It is also attractive because of its simplicity and elegance. In its basic form it can be very slow, but several modifications have been proposed that have significantly improved its performance. The first important observation concerns the strategy for restarting it. In the original algorithm proposed by Fletcher and Reeves (1964) restarting occurred every n or (n+1) steps, by setting $d_K = -g_K$. More recently Powell (1977) proposed an automatic criterion for restarting that proved to be more effective. The idea is that as long as the function resembles a quadratic, in the region where the iterates are being produced, we should continue with the iteration (2). However, it if at some step we observe a drastic non-quadratic behaviour we should restart. To measure the deviation from quadratic behaviour Powell uses the ratio

* This work was supported in part by National Science Foundation Grant No. DCR-8401903.

$$\frac{g_K^T \; g_{K-1}}{g_K^T \; g_K} \; , \tag{3}$$

which is zero for a quadratic objective function and exact line searches. If this ratio exceeds some given tolerance, say 0.2, the algorithm is restarted. In practice one finds that this strategy will lead to frequent restarts, and if the restart direction is the negative gradient, the algorithm will be very slow. Thus Powell proposes using the method of Beale (1972) that produces conjugate directions even when the initial direction is not the negative gradient. If we decide to restart we take the last search direction as the initial direction in Beale's method and proceed until a new restart is necessary. See Powell (1977) for details.

To make the restarting strategy effective, it was important to retain some information on the problem before restarting. This leads us to the second important modification of the conjugate gradient method, which consists in allowing it to store more of the information collected during the iterations. This can be done in various ways, but one can see all of these as preconditioning techniques. The algorithm of Shanno and Phua (1978) is a conjugate direction method with increased storage that incorporates Powell's restarts. It clearly out-performs the original conjugate gradient algorithm of Fletcher and Reeves. The so called variable storage methods described by Buckley (1978, 1984), Nazareth and Nocedal (1982), Gill, Murray and Wright (1981), and Nocedal (1980) can reduce the number of function evaluations even further. The appeal of variable storage methods lies in the fact that the user can decide how much storage is to be used, and thus find the most efficient algorithm for his particular application.

Another approach for improving the conjugate gradient method is by using the conic functions recently proposed by Davidon (1980). A conic function is the ratio of a quadratic over the square of an affine function, and is thus more general than a quadratic. Instead of studying algorithms that terminate only on quadratics, like the conjugate gradient method, one can derive algorithms that terminate on conics. Two such algorithms that extend the properties of the conjugate gradient method have been proposed by Davidon (1982) and Gourgeon and Nocedal (1985). These methods, however, have been tested only on conic functions thus far, and it is not known whether they will give the desired improvement.

In this paper we explore a different avenue. We will view the conjugate gradient method as a trust region algorithm, and in doing so we will describe a new formula for computing search directions. Trust region algorithms are attractive for both theoretical and practical reasons. They have good convergence properties and perform very well in practice (see for example Moré (1982)). Trust region methods are usually formulated as follows. Suppose that x_{K+1} is our current iterate and that B_{K+1} is a positive definite matrix which may or may not approximate $\nabla^2 f(x_{K+1})$. In addition suppose that Δ_{K+1} is an estimate of the size of a sphere around x_{K+1} within which the objective function is well approximated by a quadratic. Then to generate the new search direction we solve the problem

$$\min \frac{1}{2} d^T B_{K+1} d + g_{K+1}^T d \qquad \text{subject to } ||d||_2 \leq \Delta_{K+1} . \tag{4}$$

Let d_{K+1} be the solution of (4). If $||d_{K+1}||_2 < \Delta_{K+1}$ then

$$d_{K+1} = -B_{K+1}^{-1} g_{K+1} . \tag{5}$$

On the other hand, if $||d_{K+1}||_2 = \Delta_{K+1}$ then there exists a $\lambda \geq 0$ such that

$$(B_{K+1} + \lambda I) d_{K+1} = -g_{K+1} . \tag{6}$$

The value of λ is unknown; however, it is easy to show that it is unique (see Dennis and Schnabel (1983)). In current implementations of trust region methods λ is found by iteration and d_{K+1} is then computed from (6).

Let us now see how to formulate a trust region problem for the conjugate gradient method. From (2) we have

$$d_{K+1} = -(I - \frac{d_K y_K^T}{d_K^T y_K}) g_{K+1} \equiv -P_K g_K . \tag{7}$$

The matrix P_K is nonsymmetric and singular. Thus (7) is not the solution of a trust region problem of the form (4). Shanno (1978) however, has pointed out that the conjugate gradient method can be viewed as a "memoryless" BFGS algorithm, and this point of view will enable us to define a trust region problem. To describe Shanno's observation we need to look at the BFGS iteration, which is given by

$$x_{K+1} = x_K + \alpha_K d_K$$

$$d_K = -H_K\, g_K$$

$$H_{K+1} = (I - \frac{s_K\, y_K^T}{s_K^T y_K})\, H_K(I - \frac{y_K s_K^T}{s_K^T y_K}) + \frac{s_K s_K^T}{s_K^T y_K} \quad . \tag{8}$$

Here $s_K = x_{K+1} - x_K$ and, as before, $y_K = g_{K+1} - g_K$ and α_K is a steplength. Now suppose that at each step we set $H_K = I$ before computing H_{K+1} by (8). Furthermore assume that exact line searches are performed, so that $g_{K+1}^T s_K = 0$ holds for all $K \geq 0$. Then the iteration (8) coincides with (2). We conclude that the "memoryless" BFGS method using exact line searches is equivalent to the basic conjugate gradient method.

Let us therefore consider the updating formula

$$H_{K+1} = (I - \frac{s_K y_K^T}{s_K^T y_K})(I - \frac{y_K s_K^T}{s_K^T y_K}) + \frac{s_K s_K^T}{s_K^T s_K} \quad . \tag{9}$$

If we denote H_{K+1}^{-1} by B_{K+1} then one can show that

$$B_{K+1} = I + \frac{y_K y_K^T}{s_K^T y_K} - \frac{s_K s_K^T}{s_K^T s_K} \quad . \tag{10}$$

Note that the matrices $\{B_K\}$ are symmetric and positive definite provided $s_K^T y_K > 0$ for all K. We use them to formulate the trust region problem

$$\min \tfrac{1}{2}\, d^T B_{K+1} d + g_{K+1}^T d$$
$$\text{subject to } ||d||_2 \leq \Delta_K \quad . \tag{11}$$

If the solution of (11), d_{K+1}, satisfies $||d_{K+1}||_2 < \Delta_K$ then it coincides with the "memoryless" BFGS direction. It will also coincide with the conjugate gradient direction provided the line search in the previous step was exact. Under these two conditions we will recover the conjugate step. As it is undesirable to perform accurate line searches the directions generated by the trust region approach will always differ from those of the conjugate gradient method. Now suppose that the solution to (11) satisfies $||d_{K+1}||_2 = \Delta_K$ then from (6) and (10) we conclude that there exists a $\lambda \geq 0$ such that

$$[(\lambda + 1)I + \frac{y_K y_K^T}{s_K^T y_K} - \frac{s_K s_K^T}{s_K^T s_K}]d_{K+1} = -g_{K+1} \quad . \tag{12}$$

Therefore

$$(\lambda + 1)d_{K+1} + \frac{y_K^T d_{K+1}}{s_K^T y_K} y_K - \frac{s_K^T d_{K+1}}{s_K^T s_K} s_K = -g_{K+1} .$$

Note that d_{K+1} is in the span of g_{K+1}, s_K and y_K; in the conjugate gradient method d_{K+1} is in the span of g_{K+1} and s_K only. Let us now drop all the subscripts to simplify the notation. In order to compute the new search direction we need to find $\lambda \geq 0$ and d such that

$$||d||_2 = \Delta$$

and

$$(\lambda + 1)d + \frac{y^T d}{s^T y} y - \frac{s^T d}{s^T s} s = -g \quad . \tag{13}$$

Let us define $\rho = 1/y^T s$, $\sigma = 1/s^T s$. Forming the inner product of (13) with d, y, s and g we obtain

$$\begin{aligned}
(\lambda+1)d^T d + \rho(y^T d)^2 - \sigma (s^T d)^2 &= -g^T d \\
(\lambda+1)y^T d + \rho(y^T y)(y^T d) - (\sigma/\rho)(s^T d) &= -g^T y \\
(\lambda+1)s^T d + (y^T d) - s^T d &= -g^T s \\
(\lambda+1)g^T d + \rho(g^T y)(y^T d) - \sigma(s^T g)(s^T d) &= -g^T g \quad .
\end{aligned} \tag{14}$$

This is a system of 4 polynomial equations in 4 variables. Let us introduce new symbols for the variables:

$$z = (\lambda+1), \; \mu = y^T d, \; v = g^T d, \; w = s^T d \quad . \tag{15a}$$

To simplify the notation further we introduce additional symbols for some of the coefficients in (14)

$$t = \sigma/\rho, \; \alpha = \rho(y^T y), \; \beta = \rho(g^T y), \; \gamma = \sigma(g^T s), \; \eta = g^T y,$$
$$c = g^T g, \; a = \gamma/\sigma \quad . \tag{15b}$$

Observing that $d^T d = \Delta^2$ we write (14) as

$$\begin{aligned}
z\Delta^2 + \rho\mu^2 - \sigma w^2 + v &= 0 \\
z\mu + \alpha\mu - tw &= -\eta \\
zw + \mu - w &= -a \\
zv + \beta\mu - \gamma w &= -c \quad .
\end{aligned} \tag{16}$$

We will now transform (16) into a polynomial equation in one variable. Using the last three equations we will express μ,v and w in terms of z.

$$\begin{aligned}
\mu &= w(1-z)-a \\
(z+\alpha)\mu - tw &= (z+\alpha)[w(1-z) - a] - tw \\
&= w[(z+\alpha)(1-z) - t] - (z+\alpha)a \\
&= -\eta \; .
\end{aligned} \tag{17}$$

From the last equality we have

$$w = \frac{-(z+\alpha)a + \eta}{(z+\alpha)(z-1) + t} \quad . \tag{18}$$

We substitute (18) into (17) to obtain

$$\mu = \frac{(1-z)[-(z+\alpha)a + \eta]}{(z+\alpha)(z-1) + t} - a$$

$$= \frac{(z+\alpha)(z-1)a - (z-1)\eta - a(z+\alpha)(z-1) - ta}{(z+\alpha)(z-1) + t} \quad .$$

Therefore

$$\mu = \frac{-(z-1)\eta - ta}{(z+\alpha)(z-1) + t} \quad . \tag{19}$$

Now we substitute (18) and (19) into the last equation of (16)

$$v = \frac{\gamma w - \beta\mu - c}{z}$$

$$= \frac{\gamma w - \beta w(1-z) + \beta a - c}{z}$$

$$= \left[\frac{\gamma+\beta(z-1)}{z}\right]\left[\frac{-(z+\alpha)a + \eta}{(z+\alpha)(z-1) + t}\right] + \frac{\beta a - c}{z} \quad . \tag{20}$$

Finally we substitute (18), (19) and (20) into the first equation of (16)

$$z\Delta^2 + \rho\frac{[(z-1)\eta + ta]^2}{[(z+\alpha)(z-1) + t]^2} - \sigma\frac{[(z+\alpha)a - \eta]^2}{[(z+\alpha)(z-1) + t]^2}$$

$$+ \frac{[\gamma + \beta(z-1)][\eta - (z+\alpha)a]}{z[(z + \alpha)(z-1) + t]} + \frac{\beta a - c}{z} = 0$$

or

$$z^2\Delta^2[(z+\alpha)(z-1) + t]^2 + \rho z[(z-1)\eta + ta]^2 - \sigma z[(z+\alpha)a - \eta]^2$$

$$+ [(z+\alpha)(z-1)+ t][\gamma + \beta(z-1)][\eta - (z+\alpha)a] + (\beta a-c)[(z+\alpha)(z-1)+t]^2 = 0. \tag{21}$$

This is a 6th order equation in z. Since $z = \lambda+1$ and, as there is a unique $\lambda \geq 0$ that solves the trust region problem, we conclude that (21) has only one root in $[1,\infty)$. The properties of (21) have been studied in a more general setting by Hebden (1973) and Moré (1977) (see also Dennis and Schnabel (1983)). To solve (21) we can use Newton's method starting at $z = 1$. Note that since we are solving one equation in one unknown the iteration is very inexpensive. In fact, most of the computational effort involved in this approach lies in forming the coefficients of the polynomial equation (21).

Let us now collect terms in (21), and to reduce the expression, we introduce more new symbols:

$$m = \alpha-1,\ q = -\alpha+t \quad r = \eta - ta, \quad \ell = \alpha a - \eta, \quad e = \beta a - c,$$
$$h = \gamma - \beta,\ K = \beta\ell + ah . \tag{22}$$

We have that

$$[(z+\alpha)(z-1)+ t]^2 = [z^2 + (\alpha-1)z - \alpha + t]^2$$
$$= [z^2 + mz + q]^2 \tag{23a}$$

$$= z^4 + 2mz^3 + 2qz^2 + m^2z^2 + 2mzq + q^2$$
$$= z^4 + (2m)z^3 + (2q + m^2)z^2 + (2mq)z + q^2 . \tag{23b}$$

Therefore

$$z^2\Delta^2[(z+\alpha)(z-1) + t]^2 = \Delta^2z^6 + (2m\Delta^2)z^5 + (2q + m^2)\ \Delta^2z^4$$
$$+ \ (2mq)\Delta^2z^3 + \Delta^2q^2z^2 \quad . \tag{24}$$

We now expand all the other terms in (21)

$$\rho z[(z-1)\eta + ta]^2 = \rho z[\eta z - \eta + ta]^2$$
$$= \rho z[\eta z - r]^2$$
$$= \rho z[\eta^2z^2 - 2\eta rz + r^2]$$
$$= (\rho\eta^2)z^3 - (2\eta\rho r)z^2 + (\rho r^2)z , \tag{25}$$

$$-\sigma z[(z+\alpha)a - \eta]^2 = -\sigma z[za + \alpha a - \eta]^2$$
$$= -\sigma z[za + \ell]^2$$
$$= -(\sigma a^2)z^3 - (2\sigma a\ell)z^2 - (\sigma\ell^2)z . \tag{26}$$

Using (23a)

$$[(z+\alpha)(z-1) + t][\gamma + \beta(z-1)][\eta - (z+\alpha)a] =$$
$$= [z^2 + mz + q][\beta z + \gamma - \beta][-az + \eta - a\alpha]$$
$$= [z^2 + mz + q][\beta z + h][-az - \ell]$$
$$= [z^2 + mz + q][(-a\beta)z^2 - (\beta\ell + ah)z - h\ell]$$
$$= [z^2 + mz + q][(-a\beta)z^2 - Kz - h\ell]$$
$$= (-a\beta)z^4 - Kz^3 - h\ell z^2 - a\beta mz^3 - mKz^2 - mh\ell z$$
$$-a\beta qz^2 - qKz - qh\ell$$
$$= -(a\beta)z^4 - (K + a\beta m)z^3 - (mK + h\ell + a\beta q)z^2$$
$$- (qK + mh\ell)z - qh\ell \quad . \tag{27}$$

From (23b)

$$(\beta a - c)[(z+\alpha)(z-1) + t]^2 = ez^4 + (2me)z^3$$
$$+ (2q + m^2)ez^2 + (2mq)ez + q^2e \quad . \tag{28}$$

Thus substituting (23)-(28) into (21) and collecting terms we obtain

$$\Delta^2 z^6 + (2m\Delta^2)z^5 + [(2q + m^2)\Delta^2 - a\beta + e]z^4 + [2mq\Delta^2 + \rho\eta^2 - a^2\sigma - K - a\beta m + 2me]z^3 + [\Delta^2 q^2 - 2\eta\rho r - 2a\ell\sigma - mK - h\ell - a\beta q + (2q + m^2)e]z^2 + [\rho r^2 - \sigma\ell^2 - qK - mh\ell + 2mqe]z - qh\ell + q^2 e = 0 \quad . \tag{29}$$

We find $\lambda = z-1$ by solving (29) and compute d_{K+1} using (13), (18) and (19).

We have thus shown how to derive an algorithm based on the trust region problem (11) with B_K defined by (10). Another way of describing our derivation of equation (29) is by noting that the Sherman-Morrison formula can be applied twice to the matrix appearing in (12), thus expressing d_{K+1} in terms of λ. The condition $||d_{K+1}||_2 = \Delta_K$ determines λ. It is interesting to point out that variable storage methods can also be generalized in this fashion. In this case (12) will consist of a correction of rank 2m of the identify matrix, where m is the number of corrections stored. The Sherman-Morrison formula can be applied 2m times and λ is determined, as before, by the condition $||d_{K+1}|| = \Delta_K$. The formulas will be rather complicated but they can be derived using a symbolic manipulation program.

Concluding Remarks

We have presented a different view of conjugate gradient methods. The new algorithm, however, has only been described in very general terms. Many important details of implementation need to be studied. For example, it is necessary to ensure that $s_K^T y_K > 0$ so that the matrices $\{B_K\}$ remain positive definite. We suggest to use a line search to accomplish this. Note that solving the trust region problem is very inexpensive, thus we can afford the line search computation. In this respect the algorithm will differ from other trust region methods that avoid line searches altogether. It is also important to implement the algorithm so that it has an R-superlinear rate of convergence, or in other words, so that it has quadratic termination. To achieve this we may implement it in conjunction with Beale's method or a method with similar properties.

References

E.M.L. Beale (1972). A derivation of conjugate gradients, in F.A. Lootsma, ed., Numerical Methods for Nonlinear Optimization, pp. 39-43, Academic Press.

A.G. Buckley (1978). A combined conjugate-gradient quasi-Newton minimization algorithm, Math. Programming 15, 200-210.

A.G. Buckely (1984). Termination and equivalence results for conjugate gradient algorithms, Math. Programming 29, No. 1, 67-76.

W.C. Davidon (1980). Conic approximations and collinear scalings for optimizers, SIAM J. Num. Anal. 17, 268-281.

W.C. Davidon (1982). Conjugate directions for conic functions, in M.J.D. Powell, ed., Nonlinear Optimization 1981, Academic Press.

J.E. Dennis and R. Schnabel (1983). Numerical Methods for Unconstrained Optimization and Nonlinear Equations, Prentice Hall.

R. Fletcher and C. Reeves (1964). Function minimization by conjugate gradients, The Computer Journal 7, 149-154.

R. Fletcher (1970). A new approach to variable metric algorithms, Computer J. 13, 317-322.

P. Gill, W. Murray and M. Wright (1981). Practical Optimization, Academic Press.

H. Gourgeon and J. Nocedal (1985). A conic algorithm for Optimization, SIAM J. on Scientific and Statistical Computing 6, No. 2, 253-267.

M.D. Hebden (1973). An algorithm for minimization using exact second derivatives, Rept TP515, A.E.R.E., Harwell.

J.J. Moré (1977). The Levenberg-Marquardt algorithm: Implementation and theory, in G.A. Watson, ed., Numerical Analysis, Lecture Notes in Math. 630, Springer Verlag, 105-116.

J.J. Moré (1982). Recent developments in algorithms and software for trust region methods, ANL/MCS-TM-2, Argonne National Laboratory.

L. Nazareth and J. Nocedal (1982). Conjugate gradient methods with variable storage, Math. Programming 23, 326-340.

J. Nocedal (1980). Updating quasi-Newton matrices with limited storage, Math. Comp. 35, 773-782.

M.J.D. Powell (1977). Restart procedures for the conjugate gradient method, Math. Programming 12, 241-254.

D.F. Shanno (1978). Conjugate gradient methods with inexact line searches, Mathematics of Operations Research 3, 244-256.

D.F. Shanno and K. Phua (1978). A variable method subroutine for unconstrained nonlinear optimization, MIS tech. Rep. 28, University of Arizona.

AN EFFICIENT STRATEGY FOR UTILIZING A MERIT FUNCTION IN NONLINEAR PROGRAMMING ALGORITHMS

Paul T. Boggs
Center for Applied Mathematics
National Bureau of Standards
Gaithersburg, MD 20899

Jon W. Tolle
Curriculum in Operations Research
University of North Carolina
Chapel Hill, NC 27514

1. INTRODUCTION

This paper reports on the continuing effort of the authors to develop an efficient merit function for use in solving nonlinear programming problems. For the equality-constrained problem

$$\text{(NLP)} \qquad \begin{array}{rl} \text{minimize} & f(x) \\ \text{subject to} & g(x)=0 \end{array}$$

where $f : \mathbb{R}^n \to \mathbb{R}$ and $G : \mathbb{R}^n \to \mathbb{R}^m$ are smooth, the method of sequential quadratic programming (SQP) has been shown to generate good step directions for computing iterative approximations to the solution. However, the best choice of a merit, or line search, function with which to determine the appropriate step length so as to guarantee rapid convergence is still a matter of some debate. (See, for example, [ChaLPP82], [Han77], BogT84], [Pow85], [Sch83].)

The research of the authors has centered on a merit function of the form

$$(1.1) \qquad \phi_d(x) = f(x)+\bar{\lambda}(x)^T g(x)+\frac{1}{2d}g(x)^T[\nabla g(x)^T\nabla g(x)]^{-1}g(x) \ ,$$

where

$$\bar{\lambda}(x) = [\nabla g(x)^T\nabla g(x)]^{-1}\nabla g(x)^T\nabla f(x)$$

is the least squares approximation of the Lagrange multiplier vector. ϕ_d is a member of a class of exact penalty functions for NLP [BogT80] which has been shown to have certain useful properties when used in conjunction with the SQP method [BogT84].

In Section 3 of this work results are stated which show this merit function can be employed to yield a convergent algorithm if the iterations are begun close to feasibility. We therefore obtain a globally convergent algorithm if this basic procedure is modified to reduce constraint infeasibilities when the merit function fails to be reduced. In Section 4 a family of surrogate merit functions are proposed which have similar properties to ϕ_d but are more cheaply applied. The surrogate merit functions can be used together with ϕ_d to obtain a globally convergent algorithm. In Section 5 a particular implementation of the method is outlined and the results of some numerical tests are discussed. Proofs of the results stated in Sections 3 and 4 and details of the numerical experiments can be found in [BogT85].

2. THE BASIC ITERATION SCHEME

The notation and terminology used are consistent with that of [BogT84]. The results from that paper will be used as well.

For the equality-constrained nonlinear program NLP the SQP method generates a step s at an iterate x by means of the formula

$$s = -B^{-1}\{I-\nabla g(x)[\nabla g(x)^T B^{-1}\nabla g(x)]^{-1}\nabla g(x)^T B^{-1}\}\nabla f(x)$$

(2.1)

$$-B^{-1}\nabla g(x)[\nabla g(x)^T B^{-1}\nabla g(x)]^{-1} g(x) \; .$$

This formula can be derived either as a solution to a quadratic approximation to NLP (see, for example, [Han76]) or as a quasi-Newton step for solving the Karush-Kuhn-Tucker equations (see [Tap78]). In either case the matrix B is generally taken to be a positive definite approximation to the Hessian matrix (denoted $\ell_{xx}(x,\lambda)$) of the Lagrangian function

$$\ell(x,\lambda) = f(x) + \lambda^T g(x) \; .$$

It will be useful in certain situations to consider the decomposition of s into the orthogonal components:

$$s_q = Q(x)s$$

and

$$s_p = P(x)s$$

where

$$Q(x) = \nabla g(x)[\nabla g(x)^T \nabla g(x)]^{-1}\nabla g(x)^T$$

and

$$P(x) = I - Q(x) \; .$$

In this decomposition s_q and s_p represent respectively, the normal and tangential components of s with respect to the manifold

$$S_x = \{z : g(z)=g(x)\} \; .$$

For the remainder of this paper we will assume that the following conditions hold:

A1. The objective function f and the constraint function g are twice continuously differentiable.

A2. There is a unique Karush-Kuhn-Tucker point (x^*,λ^*) at which the strong second order sufficient conditions hold. In particular this implies that the matrix

$$P(x^*)\ell_{xx}(x^*,\lambda^*)P(x^*)$$

is positive definite.

A3. There is some $\bar{\eta} > 0$ such that for $\eta \leq \bar{\eta}$ the sets

$$G(\eta) = \{x: \| g(x) \| \leq \eta\}$$

are compact and there is some open set C containing $G(\bar{\eta})$ in which the matrix $\nabla g(x)^T \nabla g(x)$ is invertible.

A4. The matrices B are always chosen from a class of positive definite matrices $\mathcal{B}$ for which there exist positive constants μ and ν such that

$$\mu \| x \|^2 \leq x^T B x \leq \nu \| x \|^2$$

for all $x \varepsilon \mathbb{R}^m$ and all $B \varepsilon \mathcal{B}$.

Many of the results in this paper are valid under a weaker set of assumptions than those given above; this set has been chosen to simplify the exposition.

One important consequence of assumptions A2 and A3 is that for d and η sufficiently small the solution to NLP, x^*, is also the unique solution to

$$\min_{x \,\varepsilon\, G(\eta)} \phi_d(x) .$$

This follows from the penalty function properties of $\phi_d(x)$. (See [BogT77]).

The basic iteration scheme to which the theory in Sections 3 and 4 is devoted is described below. It is an SQP-type algorithm utilizing the merit function given in equation (1.1).

(i) Given $x^0 \varepsilon C$, $B_0 \varepsilon \mathcal{B}$, and $d > 0$, set $k = 0$.

(ii) Let s^k be the SQP step given by (2.1) when $x = x^k$ and $B = B_k$.

(iii) Choose α_k to be any positive scalar such that for all $\alpha \varepsilon (0,\alpha_k)$

$$\phi_d(x^k+\alpha s^k) < \phi_d(x^k) . \tag{2.2}$$

(iv) Set

$$x^{k+1} = x^k+\alpha_k s^k$$

and choose B_{k+1} from $\mathcal{B}$.

(v) Check the termination criteria; if they are not satisfied, set $k = k+1$ and return to (ii).

In order for this iteration scheme to be well-defined it is necessary to demonstrate that step (iii) can be carried out. In our previous paper ([BogT84], Theorems 3 and 4) the following result was established.

Lemma 2.1: Let $\bar{\eta}$ be as defined in Assumption A3. There exists a $\bar{d} > 0$ such that for each $d \in (0,\bar{d})$ there is an $\eta_d \in (0,\bar{\eta})$ for which $x \in G(\eta_d)$ and $B \in \mathcal{B}$ imply

$$\nabla\phi_d(x)^T s < 0$$

for s given by (2.1).

Thus step (iii) of the iteration scheme is always possible (in theory) provided the iterates are close enough to the feasible set and d is sufficiently small (but fixed).

3. CONVERGENCE RESULTS

That the iteration scheme described in the preceding section has certain advantageous properties was demonstrated in [BogT84]. However no convergence theory was established in that paper. In this section we show that under the assumptions of Section 2 the iteration scheme will yield a convergent sequence of iterates if the initial point is close to feasibility and the steplength is properly chosen. Thus the iteration scheme can be combined with some method which reduces infeasibility to generate a globally convergent algorithm.

Throughout this section we use the constants $\bar{\eta}$ and $\bar{d}$ from Lemma 2.1. Specifically, $\nabla g(x)$ has full rank for $x \in G(\bar{\eta})$ and for each $d \in (0,\bar{d})$ there is an $\eta_d \in (0,\bar{\eta})$ such that s is a descent direction of ϕ_d at $x \in G(\eta_d)$. The sequence $\{x^k\}$ is assumed to be generated by the basic iteration scheme.

The first lemma of this section shows that the infeasible set, $G(0)$, acts as a region of attraction for the iterates generated by our scheme. That is, once they reach a neighborhood of $G(0)$ they cannot thereafter wander far from that neighborhood.

Lemma 3.1: For every $\eta_1 \in (0,\bar{\eta})$ there exist constants $\eta_2 \in (0,\eta_1)$ and $d_1 \in (0,\bar{d})$ such that if $x^0 \in G(\eta_2)$ and $d \in (0,d_1)$ then $x^k \in G(\eta_1)$ for all k.

Sketch of proof: The proof is by contradiction; if $x^{k+1} \notin G(\eta_1)$ then ϕ_d must increase.

Next we show that the level sets of $\phi_d(x)$ are bounded by the level sets $G(\eta)$. We denote the level sets of ϕ_d by

$$H_d(\rho) = \{x : \phi_d(x) \leq \rho\}$$

and we let $H_d^*(\rho)$ be that component of $H_d(\rho)$ which contains x^*.

Lemma 3.2: There exist constants $\eta_3 \varepsilon (0,\bar{\eta})$ and $d_3 \varepsilon (0,\bar{d})$ such that for each $d \varepsilon (0,d_3)$ there exists a constant $\rho(d) > 0$ such that

$$G(\eta_3) \subset H_d^*(\rho(d)) \subset G(\bar{\eta}) .$$

Sketch of proof: For d sufficiently small, H_d^* is dominated by the last term of (1.1) and hence the appropriate constants can be found.

Lemma 2.1 shows that s is a descent direction for $\phi_d(x)$ when x is close to feasibility. The next lemma refines that result by showing that essentially $\nabla\phi_d(x)^T s$ is uniformly concave near feasibility.

Lemma 3.3: There exist positive constants $d_4 \varepsilon (0,\bar{d})$ and $\eta_4 \varepsilon (0,\bar{\eta})$ such that for any $d \varepsilon (0,d_4)$ there are positive constants ζ_d^1 and ζ_d^2 for which

$$-\zeta_d^1 \cdot \| s \|^2 \leq \nabla\phi_d(x)^T s \leq -\zeta_d^2 \cdot \| s \|^2$$

for $x \varepsilon G(\eta_4)$, $B \varepsilon \mathcal{B}$, and s given by (2.1).

Sketch of proof: The first inequality follows from well known relationships. The second is proved in three parts: the first for x feasible but $x \neq x^*$; the second for x nearly feasible but not close to x^*; and the third for x near x^*. The last case requires a delicate choice for d.

We can now combine the preceding lemmas to obtain the following theorem.

Theorem 3.4: There exist positive constants $\bar{d}$ and $\bar{\eta}$ such that if $x^0 \varepsilon G(\bar{\eta})$ and $0 < d < \bar{d}$ then the sequence, $\{x^k\}$, generated by the iteration scheme is well-defined for any choice of B_k from , remains in some compact set

$$C_d = \{x : \phi_d(x) \leq \rho(d)\} \subset G(\bar{\eta}) ,$$

where

$$G(0) \subset C_d ,$$

and satisfies

$$\frac{| \nabla\phi_d(x^k)^T s^k |}{\| s^k \|} \geq \Gamma_d \| \nabla\phi_d(x^k) \|$$

for some positive constant Γ_d.

The inequality (3.1) states that the sequence $\{s^k\}$ is "gradient related" to $\{x^k\}$ in the sense of Ortega and Rheinboldt ([OrtR70]). It follows from Theorem 14.3.2 in their book that under the hypotheses of Theorem 3.4 the sequence $\{x^k\}$ will converge to x^* provided that

$$(3.2) \qquad \frac{\nabla\phi_d(x^k)^T s^k}{\| s^k \|} \to 0 .$$

There are a number of conditions that can be imposed on the choice of the step length parameter α_k to ensure that (3.2) holds. For example, if α_k is the first local minimum of $\phi_d(x^k+\alpha s^k)$, $\alpha \geq 0$, then (3.2) holds. Another popular method for choosing the α_k which leads to convergence is to satisfy the set of criteria due to Armijo and Goldstein. For a discussion of these and other possibilities the reader is referred to Section 14.2 of the aforementioned book.

As a consequence of the above it is seen that the iteration scheme proposed in Section 2 can be implemented to give a convergent algorithm provided that x^0 is chosen close enough to feasibility and the step length parameter is chosen so that (2.2) and (3.2) are satisfied. Thus, if this algorithm is combined with a procedure to reduce infeasibility, a globally convergent procedure will result.

4. A MODIFIED MERIT FUNCTION

It has been shown in [BogT84] and in the previous section that the merit function $\phi_d(x)$ has many desirable properties when employed in conjunction with an SQP-type algorithm. There are, however, two problems associated with the practical implementation of a procedure based on this function.

The first difficulty is due to the presence of the parameter d which must be specified in an appropriate manner for any particular application. This difficulty seems to be a generic one in nonlinear programming. Most, if not all, algorithms for solving NLP (with nonlinear constraints) employ a parameter in one way or another. In Section 5 we suggest a means of choosing the parameter and comment on the sensitivity of the algorithm to that choice.

The second disadvantage in the direct use of the merit function $\phi_d(x)$, and the one addressed in this section, is the amount of effort required to evaluate $\phi_d(x)$ at a tentative new iterate. That is, given a value of d and a current iterate x^k, the iterative scheme generates a direction s^k and a new point x where

$$x = x^k + \alpha s^k$$

for some $\alpha > 0$. x may be accepted as x^{k+1} or rejected according to some specified test on the value of $\phi_d(x)$. (See Section 3.) If x is rejected then a new α is chosen and the test is repeated. Since each test requires the evaluation of $\phi_d(x)$, and $\phi_d(x)$ contains the derivatives of the objective and constraint functions, the performance of the algorithm can be seriously degraded if very many tests are required at each step.

In the formula for $\phi_d(x)$ given in (1.1) it is the gradient terms that involve the most work to compute. Thus, in order to avoid time-consuming evaluations of these derivatives, we will use the modified merit function, $\phi_d^k(x)$, which is the function defined by the current iterate x^k:

$$\phi_d^k(x) = \frac{1}{d} g(x)^T A_k g(x) + \ell(x, \lambda_k)$$

where

$$A_k = [\nabla g(x^k)^T \nabla g(x^k)]^{-1}$$

and

$$\lambda_k = (\nabla f(x^k)^T \nabla g(x^k) A_k)^T .$$

The manner in which this function is used is described in Section 5.

Our first result shows that s^k is a descent direction for $\phi_d^k(x)$ at $x = x^k$.

<u>Theorem 4.1</u>: Suppose $x^k \in C$ and $B_k \in \mathcal{B}$. Then there is a constant $\bar{d} > 0$ such that for $0 < d < \bar{d}$

$$\nabla \phi_d^k(x^k)^T s^k < 0 .$$

<u>Sketch of proof</u>: The proof uses an easily derived formula for $\nabla \phi_d^k(x^k)$ and a careful choice of d.

Note that this result is stronger than the corresponding one for $\phi_d(x)$ (Lemma (2.1)) in one sense. That is, s^k is a descent direction regardless of whether or not x^k is close to feasibility. Of course, $\phi_d^k(x)$ changes from iterate to iterate and hence no decrease in a single merit function is obtained. As explained in the next section, we use $\phi_d^k(x)$ as a replacement for the true merit function $\phi_d(x)$ only at x^k, i.e., we use ϕ_d^k for the line search but ϕ_d to monitor the iterations and hence ensure convergence.

In [BogT84] it was shown that

$$\phi_d(x^k+s^k) < \phi_d(x^k)$$

provided the sequence of iterates $\{x^k\}$ is converging to x^* q-superlinearly. While this result does not imply that superlinear convergence holds for an algorithm which uses ϕ_d as a merit function (indeed, theoretical q-superlinear convergence has not been demonstrated in general for any SQP method) it suggests that this choice of merit function will not impede superlinear convergence when it occurs, e.g., in convex problems.

The second theorem of this section demonstrates that the modified merit function ϕ_d^k also has the property, i.e., a unit step length will always be allowed if the sequence of iterates is converging q-superlinearly.

<u>Theorem 4.2</u>: Let $\{x^k\}$ be generated by an implementation of the iteration scheme and suppose that the sequence converges q-superlinearly. Then there exists a $\bar{d} > 0$ such that for each $d \in (0,\bar{d})$ there is a positive integer $J(d)$ such that for all $k \geqq J(d)$

$$\phi_d^k(x^k+\alpha s^k) < \phi_d^k(x^k)$$

whenever $0 < \alpha \leqq 1$.

<u>Sketch of proof</u>: The proof uses a Taylor series expansion of $\phi_d^k(x^k+s^k)$. The result then follows from the use of the characterization of q-superlinear convergence contained in [BogTW82].

5. THE ALGORITHM AND NUMERICAL RESULTS

The discussion in the previous sections allows us to state an algorithm which is globally convergent, efficient, relatively simple, and one which has very few arbitrary parameters. The basic procedure is to use ϕ_d^k as defined in Section 4 as the local merit function. Without some modification, this would, of course, sacrifice global convergence. The modification takes the form of a monitor routine which, in accordance with Section 3, keeps track of the values of ϕ_d, adjusts d if appropriate, and arranges for infeasibility reduction if difficulties arise, i.e., if ϕ_d is not being reduced. We first discuss the use of ϕ_d^k as a local merit function and then describe the monitor (global) routine.

For description of the implementation, it is necessary to separate the components of ϕ_d and ϕ_d^k. In accordance with the notation of Sections 3 and 4 let

$$\phi_1(x) = g(x)^1[\nabla g(x)^1 \nabla g(x)]^{-1} g(x)$$

$$\phi_2(x) = \ell(x,\bar{\lambda}(x))$$

$$\phi_1^k(x) = g(x)^T A_k(x) g(x)$$

$$\phi_2^k(x) = \ell(x,\lambda_k) .$$

Note that ϕ_1 and ϕ_1^k are minimized at feasibility.

We use ϕ_d^k as a normal line search function. (Recall that this only requires evaluation of f and g, but no gradients.) This is satisfactory if d is small enough (cf. Theorem 4.1) and if good global progress is being made. If either d is too large or global considerations dictate, we try to reduce ϕ_1, which can always be done when $\phi_1^k \neq 0$. This procedure has the effect of reducing constraint infeasibilities. The value of d is not adjusted in this routine; any adjustment to d must be done with global convergence in mind and hence such decisions are made in the monitor routine described next.

The above procedure will always generate steps which either reduce ϕ_d^k or ϕ_1^k, but not necessarily ϕ_d. After such a step is determined, the gradients can be evaluated at the new point, x^+, and $\phi_d(x^+)$ can be computed. This evaluation of ϕ_d is done in the monitor routine.

The monitor routine has three parts. On its first call, the monitor initializes ϕ_1, ϕ_2, d, and ϕ_d. The second part is employed when ϕ_d^k is being satisfactorily reduced by the local routine. In this case, it evaluates ϕ_d and keeps track of the smallest value of ϕ_d so far encountered as well as the number of iterations since ϕ_d has been reduced. If reduction has not occurred over several iterations, it informs the local routine to begin reducing ϕ_1^k. The third part is invoked while the local routine is reducing ϕ_1^k. It checks for reduction of ϕ_1 only. In this case, it will adjust d to a lower value. The value of d is initially set to one since that value has worked well in practice. If d needs to be recomputed, it is set to the minimum of

$$\{.9d,\ .9^* \text{abs}(\phi_1^L - \phi_1)/\text{abs}(\phi_2^L - \phi_2)\}$$

where ϕ_1^L and ϕ_2^L correspond to that value of x which produced the lowest value of ϕ_d so far encountered. This choice has performed well on our tests. We note here that when ϕ_d^k is not easily reduced, it is often due to a poor value of d . Thus, it often occurs that when the monitor is called to check the reduction of ϕ_1 , it obtains reduction immediately and computes a new value of d . We have also implemented an option to keep d fixed at its initial value. Thus we were able to assess the performance of this automatic adjustment strategy versus keeping d fixed.

The procedure can fail when an iterate is feasible (or very nearly so) and d is very small. In this case, an exceedingly short step may be required to reduce ϕ_d^k and this may cause a line search failure. When this occurs the monitor restores the value of x corresponding to ϕ_d^L and then attempts to continue from there. If this is impossible, the routine terminates. At this point, the value of d could be increased by some sort of ad hoc procedure.

This algorithm has been coded and tried on the test problems in [BogT84]. The results show that our algorithm is not overly sensitive to the choice of d . In fact, the strategy of fixing d and never adjusting it performed almost as well as the automatic adjustment procedure. We often observed apparent q-superlinear convergence. One failure of the type mentioned above was encountered, but restarting the algorithm where the failure occurred (equivalent to increasing d) eventually led to convergence. More details of the testing are reported in [BogT85].

REFERENCES

[BogT77] P. BOGGS and J. TOLLE, "A two parameter multiplier function with exact penalty functions," Tech. Rep. No. 77-1, Curriculum in Operations Research and Systems Analysis, Univ. North Carolina, Chapel Hill, 1977.

[BogT80] _____, "Augmented Lagrangians which are quadratic in the multiplier," J. Optim. Theory Appl., 31 (1980), pp. 17-26.

[BogT84] _____, "A family of descent functions for constrained optimization," SIAM J. Numer. Anal., 21 (1984), pp. 1146-1161.

[BogT85] _____, "The implementation and testing of a merit function for constrained optimization problems," Tech. Rep. No. 85-5, Curriculum in Operations Research and Systems Analysis, Univ. North Carolina, Chapel Hill, 1985.

[BogTW82] BOGGS, P., J. TOLLE, and P. WANG (1982), "On the local convergence of quasi-Newton methods for constrained optimization," SIAM J. Control and Optimization, 20, 161-171.

[ChalPP82] R. CHAMBERLAIN, C. LEMARECHAL, H.C. PEDERSEN and M. POWELL, "The watchdog technique for forcing convergence in algorithms for constrained optimization," Mathematical Programming Study, 16 (1982), pp. 1-17.

[Han76] S. HAN, "Superlinearly convergent variable metric algorithms for general nonlinear programming problems," Math. Programming, 11 (1976), pp. 263-82.

[Han77] ______, "A globally convergent method for nonlinear programming," J. Optim. Theory Appl., 22 (1977), pp. 297-309.

[OrtR70] J. ORTEGA and W. RHEINBOLDT, <u>Iterative Solutions of Nonlinear Equations in Several Variables</u>, Academic Press, New York, 1970.

[Pow85] M. POWELL, "The performance of two subroutines for constrained optimization on some difficult test problems," Proceedings of the SIAM Conference on Numerical Optimization, (to appear).

[Sch83] K. SCHITTKOWSKI, "On the convergence of a sequential quadratic programming method with an augmented Lagrangian line search function," Math Operationsforsch. U. Statist., Ser. Optimization, 14 (1983), pp. 197-216.

[Tap78] R. TAPIA, "Quasi-Newton methods for equality constrained optimization: equivalence of existing methods and a new implementation," Nonlinear Programming 3, O. Mangasarian, R. Meyer and S. Robinson, eds., Academic Press, New York, 1978, pp. 125-64.

RATES OF CONVERGENCE FOR SECANT METHODS ON NONLINEAR PROBLEMS IN HILBERT SPACE

Andreas Griewank*
Southern Methodist University
Dallas, Texas 75275/USA

Abstract

The numerical performance of iterative methods applied to discretized operator equations may depend strongly on their theoretical rate of convergence on the underlying problem $g(x) = 0$ in Hilbert space. It is found that the usual invertibility and smoothness assumptions on the Frechet derivative $g'(x)$ are sufficient for local and linear but not necessarily superlinear convergence of secant methods. For both Broyden's Method and Variable Metric Methods it is shown that the asymptotic rate of convergence depends on the essential norm of the discrepancy D_0 between the Frechet derivative g' at the solution x_* and its initial approximation B_0. In particular one obtains local and Q-superlinear convergence if D_0 is compact which can be ensured in the case of mildly nonlinear problems where $g'(x_*)$ is known up to a compact perturbation.

1. Introduction

To motivate the analysis of secant methods in Hilbert spaces we consider the behaviour of Newton's and Broyden's method on two particular operator equations in the space ℓ^2 of square summable sequences $x = \langle (x)_i \rangle_{i=1}^{\infty}$.

The diagonal operator

$$g(x) = \left\langle \frac{(x)_i - 1/i}{i(1+|i(x)_i - 1|)} \right\rangle_{i=1}^{\infty} : \ell^2 \to \ell^2$$

has the unique root $x_* = \langle 1/i \rangle_{i=1}^{\infty} \in \ell^2$. Its Fréchet-derivative

$$g'(x) = \mathrm{diag}\left\langle \frac{1}{i(1+|i(x)_i - 1|)^2} \right\rangle_{i=1}^{\infty}$$

Keywords: Secant Methods, Variational Characterization of Eigenvalues, Compact Operators

Running Head: Secant Methods in Hilbert Space.

* This work was supported by NSF grant DMS-8401023.

is bounded and varies Lipschitz continuously in x. The Newton iterates $x_k = \langle (x_k)_i \rangle_{i=1}^{\infty}$ generated from some starting point $x_0 \in \ell^2$ satisfy the recurrence

$$(x_{k+1} - x_*)_i = -i \cdot (x_k - x_*)_i \cdot |x_k - x_*|_i \ .$$

Consequently the i-th component $(x_k)_i$ converges quadratically to $(x_*)_i = 1/i$ if and only if $|(x_0)_i - 1/i| < 1/i$.

Combining the first n equations we find that Newton's method exhibits local and quadratic convergence on any of the truncated systems

$$P_n\, g(x) = 0 \ , \qquad x = P_n\, x \tag{1.1}$$

where for all $z = \langle (z)_i \rangle_{i=1}^{\infty} \in \ell^2$

$$P_n\, z = \langle (z)_1, (z)_2, \ldots, (z)_n, 0, 0, \ldots\ 0\ \ldots \rangle \ . \tag{1.2}$$

The same is true for any other Ritz-Galerkin discretization, i.e. P_n replaced by any orthogonal projector $P : \ell^2 \to \ell^2$ of finite rank. Thus all seems well from a "practical" point of view since all computer calculations are restricted to a finite number of "real" variables anyway.

To demonstrate that this assessment is unduly optimistic let us consider the Newton iterates $P_n\, x_k$ generated on (1.1) from the particular starting point $P_n\, x_0$ with $x_0 = \langle 1/i^2 \rangle_{i=1}^{\infty}$. One can easily check by induction that for all $k \geq 0$

$$(x_k)_i = \frac{1}{i}[1 - (-1)^k (1 - 1/i)^{2^k}]$$

which implies by elementary but tedious arguments that in the 2-norm

$$\| P_n\, x_k - x_* \| > \| x_k - x_* \| \geq \tfrac{1}{2}(1/\sqrt{2})^k \ .$$

This means that the distance between the iterates $P_n\, x_k$ and the actual solution x_* is always greater than $\| x_k - x_* \|$ which declines only linearly. In other words the quadratic rate of convergence on the finite dimensional problems (1.1) is a mere artefact of the discretization because the underlying problem in ℓ^2 is not sufficiently regular. In our example the inverse Jacobian

$$g'(x_*)^{-1} = \mathrm{diag}\langle\, i\, \rangle_{i=1}^{\infty}$$

is unbounded and Newton's method may diverge even if $\| x_0 - x_* \|$ is arbitrarily small. Naturally Newton's method does converge quadratically on both the original problem and suitable discretization, if

$g'(x)$ has a bounded inverse and varies Lipschitz-continuously in x [1]. Whenever an operator equation $g(x) = 0$ violates either condition we must expect slower convergence of Newton's method, even though it might in theory converge quadratically on certain discretizations.

A similar situation may arise for so called quasi-Newton or secant methods [6]. These Newton-like methods can achieve rapid convergence while requiring neither the mathematical derivation nor the repeated evaluation and factorization of derivative matrices. In the finite-dimensional case invertibility and Hölder continuity of the Jacobian $g'(x)$ at a root $x_* \in g^{-1}(0)$ are sufficient for local and Q-superlinear convergence, i.e.

$$\lim \|x_{k+1} - x_*\| / \|x_k - x_*\| = 0 .$$

This result holds in particular for the full step Broyden iteration defined by

$$x_{k+1} - x_k = s_k \equiv -B_k^{-1}\, g(x_k) \tag{1.3}$$

and

$$B_{k+1} - B_k = (y_k - B_k\, s_k)\; s_k^T / s_k^T\, s_k \tag{1.4}$$

where the B_k are bounded linear operators and

$$y_k \equiv g(x_{k+1}) - g(x_k) .$$

Here and throughout the paper s^T denotes the linear functional associated with each element of a Hilbert space by the Riesz representation theorem [21]. On ℓ_2 we can no longer expect superlinear convergence under the usual initial condition that $x_0 - x_*$ and $B_0 - g'(x_*)$ be sufficiently small, in the vector and induced operator norm respectively.

To see this we consider the linear operator

$$g(x) \equiv B_*(x - x_*) : \ell^2 \to \ell^2$$

where $x_* = \langle 1/i \rangle_{i=1}^{\infty}$ and the only nonzero entries of $B_* - I$ are the subdiagonal elements $1 > \alpha_1 \geq \alpha_2 \ldots \geq \alpha_j \geq \alpha_{j+1} \geq \ldots \geq \alpha_* \equiv \lim \alpha_j \geq 0$. Since $\|B_* - I\| = \alpha_1 < 1$ the infinite matrix B_* has a bounded inverse whose first column is formed by the vector $b = \langle \beta_j \rangle_{j=1}^{\infty}$ with $\beta_1 = 1$ and $\beta_{j+1} = -\beta_j \times \alpha_j$ for $j \geq 1$. Starting from $x_0 \equiv x_* + b$ with $B_0 \equiv I$ it follows by induction that for all $k \geq 0$

$$(x_k - x_*)^T = \langle 0, 0, \ldots 0, \beta_{k+1}, \beta_{k+2}, \ldots \rangle \tag{1.5}$$

$$g_k^T \equiv g(x_k)^T = \langle 0, 0, \ldots 0, \beta_{k+1}, 0 \quad , \ldots \rangle \tag{1.6}$$

and

$$B_k - I = \text{zero except for the subdiagonal}$$

$$\langle \alpha_1, \alpha_2, \ldots, \alpha_k, 0, 0, 0 \ldots \rangle .$$

Thus the approximating Jacobians B_k do actually converge to the exact derivative operator B_* . However the relations (1.5) and (1.6) hold even if B_k is kept constant equal to $B_0 = I$ so that updating according to (1.4) does not really help at all. In any case we have

$$\|g_k\| / \|g_{k-1}\| = |\beta_{k+1}| / |\beta_k| = \alpha_k$$

so that the reduction of the residual at the k-th step is directly related to the k-th largest singular value α_k of $D_0 = B_0 - B_*$. Obviously we have Q-superlinear convergence if and only if $\alpha_* = 0$ which requires that D_0 be compact. Otherwise we have only linear convergence such that for all k and sufficiently large n

$$\|P_n x_k - x_*\| \geq \|x_k - x_*\| \geq \alpha_*^k .$$

Here P_n is again the truncation operator defined in (1.2) and $P_n x_k$ represents the k-th Broyden iterate generated on the system $P_n g(x) = 0$, $P_n x = x$ from the starting point $P_n x_0$ with $B_0 = I$. For this and any other Ritz-Galerkin discretization it follows by a result of Burmeister and Gay that the iteration terminates after finitely many steps at an iterate $P_n x_k$ with $P_n g(P_n x_k) = 0$. However as in the case of Newton's method on a singular problem, this discretization effect only disguises the slow progress towards the actual solution x_*.

In order to achieve genuine superlinear convergence we must require that the discrepancy D_0 between $g'(x_*)$ and its initial approximation be compact. In practice this condition can be met for so called mildly nonlinear (or essentially linear) problems, where $g'(x_*)$ is a compact perturbation of an a priori known linear operator B_0 . For example one might set $B_0 = I$ in case of a weakly singular integral equation

$$g(x) \equiv x(t) + \int_0^1 \frac{K(t,\tau,x(\tau))}{|t-\tau|^e} d\tau - h(t) = 0 .$$

The Urison operator g maps $L^2[0,1]$ onto itself provided

$$K \in C([0,1]^2 \times \mathbb{R}) , \quad 0 < e < 1 , \quad h \in L^2[0,1]$$

and K has a bounded x-derivative

$$|K_x(t,\tau,x)| \leq c \quad \text{on} \quad [0,1]^2 \times \mathbb{R} .$$

Then the Frechét derivative defined by

$$g'(x)v = v(t) + \int_0^1 \frac{K_x(t,\tau,x(\tau))}{|t-\tau|^e} v(\tau)d\tau \quad \text{for} \quad v \in L^2[0,1]$$

is everywhere a compact perturbation of the identity operator $I \cdot v = v$. Moreover if $e < \frac{1}{2}$ then $D_0 = I - g'(x_*)$ belongs to the so-called Hilbert-Schmidt class of compact operators whose singular values are square summable so that

$$\|D_0\|_F^2 \equiv \text{Trace}(D_0^T D_0) < \infty .$$

Since the same is true for any other integral equation with square integrable kernel, there is a wide class of problems where B_0 can be chosen such that $D_0 = B_0 - B_*$ has a finite Frobenius norm $\|D_0\|_F$. Under these conditions one can establish the local and Q-superlinear convergence of most secant methods by simply transcribing the classical theory [3] into Hilbert space notation. For Broyden's method this was done be Sachs [14].

In this paper we obtain the same results assuming only compactness of D_0, boundedness of $g'(x_*)^{-1}$ and a certain bound on the truncation error $g(x) - g'(x_*)(x - x_*)$. Since the Frobenius norm of all discrepancies $D_k = B_k - g'(x_*)$ may be infinite we will track instead the individual eigenvalue $\lambda_j(A_k)$ of $A_k = D_k^T D_k$. In particular we observe that at every step the Broyden update reduces each singular value of D_k up to higher order terms. As it turns out the asymptotic convergence rate of Variable Metric Methods for optimization can be studied in the same framework as Broyden's method. This analysis is carried out in the central Section 3, in which we draw heavily on techniques developed by R. Winther in his doctoral dissertation [20].

He assumed linearity of g and, in the optimization case, an exact line search in combination with the Variable Metric Update. It is well known that the resulting iterates are identical to those generated by the conjugate gradient method which was analysed by Hayes [11] and Daniel [4] in a Hilbert space setting. Stoer [16] and Powell [12] have given examples where the Variable Metric Method with exact line search achieves only linear convergence as the Hessian has a continuous spectrum. Winther [19] asserted for the same method and equivalently conjugate gradients that R-superlinear convergence is obtained if the spectrum has only finitely many cluster points. It is not clear whether a similar result can be obtained for methods without line search. Another question that we will not address is whether the approximations B_k converge in some sense to a limiting operator. Ren-Pu and Powell [13] as well as Stoer [17] have recently shown that this is true for Variable Metric Methods in Euclidean spaces.

In the following Section 2 we discuss the essential characteristics of the problems and methods under consideration. In particular we establish local and linear convergence as a basis for the asymptotic analysis in Section 3. The paper concludes with a brief summary in Section 4.

2. Local and Q-linear Convergence

With X and Y separable real Hilbert spaces we consider a possibly nonlinear operator

$$g : \mathcal{D} \subset X \to Y$$

on an open domain $\mathcal{D}$ containing some root

$$x_* \in g^{-1}(0) = \{x \in \mathcal{D} | g(x) = 0\} .$$

Moreover we assume that g has a Frechét derivative

$$B_* \in \mathcal{B}(X,Y) \quad \text{with} \quad B_*^{-1} \in \mathcal{B}(Y,X)$$

such that for some nondecreasing function $\gamma : [0,\infty) \to \mathbb{R}$ and all $x \in \mathcal{D}$

$$\|g(x) - B_*(x - x_*)\| \leq \|x - x_*\| \gamma(\|x - x_*\|) \tag{2.1}$$

as well as

$$\lim_{\rho_0 \to 0} \int_0^{\rho_0} \frac{1}{\rho} \gamma(\rho) d\rho = 0 . \tag{2.2}$$

These conditions hold with $\gamma(\rho)$ some multiple of a positive power ρ^p if g has a Hölder continuous Frechét-derivative on some neighbourhood of x_* . However our assumptions do not even require continuity of g near x_* as they are satisfied by the slightly contrived example

$$g(x) = x + \begin{cases} 0 & \text{if } x \text{ is rational} \\ x(ln\|x\|)^{-2} & \text{otherwise} \end{cases}$$

with $\tau(\rho) = (ln\rho)^{-2}$ and $X = Y = \ell^2$ or $X = Y = \mathbb{R}$. Whereas Newton's method can not even be defined on such a function we shall find that secant methods are able to achieve local and Q-superlinear convergence.

To solve the operator equation $g(x) = 0$ we apply a full-step quasi-Newton iteration of the form

$$x_{k+1} - x_k = s_k = -B_k^{-1} g_k \tag{2.3}$$

and

$$B_{k+1} - B_k = U(B_k, s_k, y_k) \qquad \text{with} \quad y_k = g_{k+1} - g_k \tag{2.4}$$

or

$$B_{k+1} - B_k = V(B_k, s_k, y_k, \phi_k) \quad \text{with} \quad \phi_k \in [0,1] \ . \tag{2.5}$$

Here $U : \mathcal{B}(X,Y) \times X \times Y \to \mathcal{B}(X,Y)$ denotes the Broyden [2] update function

$$U(B,s,y) \equiv (y - Bs)s^T/s^Ts \tag{2.6}$$

and $V : \mathcal{B}(X,X) \times X \times X \times [0,1] \to \mathcal{B}(X,X)$ the Variable Metric update function

$$V(B,s,y,\phi) = yy^T/y^Ts - Bss^TB/s^TBs + \phi ww^T \tag{2.7}$$

with

$$w \equiv \sqrt{s^TBs}\ (y/y^Ts - Bs/s^TBs) \in X \ .$$

Whereas the Broyden formula U is always applicable the Variable Metric update V was designed for the minimisation case where $Y = X$ and B_* is a priori known to be selfadjoint and positive definite. One can easily check that V maintains both these properties for the approximations B_k provided

$$y_k{}^Ts_k > 0 \quad \text{for all} \quad k \geq 0 \ .$$

Under our weak continuity assumptions this curvature condition is not automatically satisfied, even if x_k and x_{k+1} are very close to x_*.

It is well known that the secant methods defined by equations (2.3), (2.4) or (2.3), (2.5) are invariant with respect to bicontinuous linear transformations on the range Y or the domain X respectively, provided B_0 is adjusted accordingly. Thus we can premultiply the original equation $g(x) = 0$ by B_*^{-1} in the nonsymmetric case and transform the variable vector x by the positive definite root $B_*^{-1/2}$ of B_*^{-1} in the optimisation case. Most of our conditions and assertions, including in particular the compactness of D_0 and the concept of Q-superlinear convergence, apply equivalently to the original and transformed problem. Therefore we will assume from now on that $B_* = I$ or equivalently that

$$\|g(x) - (x - x_*)\| \leq \|x - x_*\|\gamma(\|x - x_*\|) \tag{2.8}$$

where $\gamma(\rho)$ satisfies (2.2). In preparation for the local and linear convergence theorem at the end of this section we prove the following lemma.

Lemma 2.1. *Let* x *and* $x + s$ *be any two points in* $\mathcal{D}$ *such that for some* $q < 1$

$$\|x + s - x_*\| \leq q\|x - x_*\| \quad \text{and} \quad \gamma(\|x - x_*\|) \leq \tfrac{1}{3}(1 - q) \ . \tag{2.9}$$

Then $y \equiv g(x+s) - g(x)$ *satisfies*

$$|y^Ts/s^Ts - 1| \le \|y-s\|/\|s\| \le 2\gamma(\|x-x_*\|)/(1-q) \le 2/3 \; . \qquad (2.10)$$

Moreover we have for all $B \in \mathcal{B}(X,Y)$

$$\|U(B,s,y) - U(B,s,s)\| \le 2\gamma(\|x-x_*\|)/(1-q) \qquad (2.11)$$

and for all positive definite $B = B^T \in \mathcal{B}(X,X)$

$$\|V(B,s,y,\phi) - V(B,s,s,\phi)\| \le 3(1+8\|B\|)\gamma(\|x-x_*\|)/(1-q) \qquad (2.12)$$

where $\phi \in [0,1]$ *is arbitrary.*

Proof. Firstly we note that by the inverse triangle inequality

$$\|s\| \ge \|x-x_*\| - \|x-x_*+s\| \ge (1-q)\|x-x_*\| \; .$$

Now it follows from (2.8) and the monotonicity of γ that

$$\begin{aligned} |y^Ts/s^Ts - 1| &= |s^T(y-s)|/s^Ts \le \|y-s\|/\|s\| \\ &\le \|g(x+s) - (x+s-x_*) - g(x) - (x-x_*)\| \\ &\le \|x+s-x_*\|\gamma(\|x+s-x_*\|) + \|x-x_*\|\gamma(\|x-x_*\|) \\ &\le (1+q)\|x-x_*\|\gamma(\|x-x_*\|) \le 2\gamma(\|x-x_*\|)\|s\|/(1-q) \le \|s\|2/3 \; . \end{aligned}$$

Hence we have established (2.10) which implies (2.11) as

$$\|U(B,s,y) - U(B,s,s)\| = \|(y-s)s^T/s^Ts\| = \|y-s\|/\|s\| \; .$$

An elementary examination shows that

$$\sin(s,y) \equiv [1 - (y^Ts)^2/(\|s\|\|y\|)^2]^{\frac{1}{2}} \le \|y-s\|/\|s\|$$

and consequently by (2.10)

$$\cos(s,y) \equiv y^Ts/(\|s\|\|y\|) \ge \sqrt{1-4/9} > 2/3 \; .$$

Abbreviating $z \equiv (y\|s\|/y^Ts - s/\|s\|)$ we find furthermore that

$$\begin{aligned} \|z\| &= \tan(s,y) = \sin(s,y)/\cos(s,y) \\ &\le 1.5\|y-s\|/\|s\| \le 3\gamma(\|x-x_*\|)/(1-q) \le 1 \; . \end{aligned} \qquad (2.13)$$

From the definition (2.7) one obtains the rather lengthy expression

$$\begin{aligned} V(B,s,y,\phi) - V(B,s,s,\phi) &= yy^T/y^Ts - ss^T/s^Ts + \phi[z(z+2s/\|s\|)^T \\ &\quad + \phi(z+2s/\|s\|)z^T]s^TBs/s^Ts + \phi[Bsz^T + zs^TB]/\|s\| \; . \end{aligned} \qquad (2.14)$$

Since the symmetric rank 2 matrix $yy^T/y^Ts - ss^T/s^Ts$ has the null vector $s \ne 0$ its norm equals the absolute value of its trace

$$\begin{aligned} \|yy^T/y^Ts - ss^T/s^Ts\| &= |\|y\|^2/y^Ts - 1| = |(y-s)^Ty/y^Ts| \\ &\le \|y-s\|/[\|s\|\cdot\cos(s,y)] \le 3\gamma(\|x-x_*\|)/(1-q) \; . \end{aligned}$$

Thus we obtain from (2.13) and (2.14) by the triangle inequality

$$\|V(B,s,y,\phi) - V(B,s,s,\phi)\| \le 3\gamma(\|x-x_*\|)/(1-q) + 2\phi\|B\|[\|z\|(\|z\|+2) + \|z\|]$$

$$\le 3(1+8\|B\|)\gamma(\|x-x_*\|)/(1-q)$$

which completes the proof of lemma 2.1. □

As we shall see in the next section the two updates satisfy for all $s \in X$ and $B \in \mathcal{B}(X,Y)$

$$\|B + U(B,s,s) - I\| \le \|B - I\| \tag{2.15}$$

and for any $\phi \in [0,1]$

$$\|B + V(B,s,s,\phi) - I\| \le \|B - I\| \tag{2.16}$$

where B is assumed to be positive definite in the second case. In combination with (2.11), (2.12) these inequalities ensure bounded deterioration with respect to the induced operator norm. This desirable property was already mentioned in [7] and it facilitates the proof of the following local convergence result.

<u>Theorem 2.2</u>. *Let* $B_0 \in \mathcal{B}(X,Y)$ *be chosen such that*

$$\|D_0\| < \tfrac{1}{2} \quad \textit{with} \quad D_k \equiv B_k - I \quad \textit{for} \quad k \ge 0 .$$

Then there exists a $\rho_0 > 0$ *such that the iterates* x_k *generated by Broyden's method* (2.3), (2.4) *from some* x_0 *with* $\|x_0 - x_*\| \le \rho_0$ *satisfy for all* $k \ge 0$

$$\|x_{k+1} - x_*\|/\|x_k - x_*\| \le q \equiv \tfrac{1}{2} + \|D_0\| < 1 \tag{2.17}$$

and

$$\overline{\lim}\ \|g_{k+1}\|/\|g_k\| \le \overline{\lim}\ 2\|D_k\ s_k\|/\|s_k\| . \tag{2.18}$$

The same is true for the Variable Metric Method (2.3), (2.5) *with* B_0 *selfadjoint and an arbitrary sequence* $\langle \phi_k \rangle_{k=0}^{\infty} \subset [0,1]$. *Moreover we have in either case*

$$\|D_k\| \le \tfrac{1}{2}q < \tfrac{1}{2} \quad \textit{for} \quad k \ge 0 \tag{2.19}$$

and

$$\sum_{k=0}^{\infty} \gamma_k < \infty \quad \textit{with} \quad \gamma_k \equiv \gamma(\|x_k - x_*\|) . \tag{2.20}$$

<u>Proof</u>. Because of our assumption (2.2) we can define $\rho_0 > 0$ as some positive number for which

$$\hat{\gamma}(\rho_0) \equiv \gamma(\rho_0) + \frac{1}{|\ln q|} \int_0^{\rho_0} \frac{1}{\rho}\, \gamma(\rho)\, d\rho \le \frac{1}{80}\,(1-q)^2$$

which implies in particular that

$$\gamma_0 \equiv \gamma(\rho_0) \le \frac{1}{80}\,(1-q)^2 \le \tfrac{1}{2}q(1-q) . \tag{2.21}$$

Now suppose the assertions (2.17) and (2.19) hold for $k = 0,1,2..j-1$

with $j \geq 1$. Then it follows from the monotonicity of $\gamma(\rho)$ by geometrical arguments that

$$\sum_{k=0}^{j-1} \gamma(\|x_k - x_*\|) \leq \sum_{k=0}^{j-1} \gamma(\rho_0 q^k) \leq \gamma(\rho_0) + \int_0^\infty \gamma(\rho_0 q^\kappa) d\kappa \leq \hat{\gamma}(\rho_0) .$$

Using (2.11) and (2.15) we derive for $D_j = B_j - I$

$$\begin{aligned} \|D_j\| &= \|D_{j-1} + U(B_{j-1}, s_{j-1}, y_{j-1}) - I\| \\ &\leq \|D_{j-1} + U(B_{j-1}, s_{j-1}, s_{j-1})\| + \|U(B_{j-1}, s_{j-1}, y_{j-1}) \\ &\qquad - U(B_{j-1}, s_{j-1}, s_{j-1})\| \\ &\leq \|D_{j-1}\| + 2\gamma(\|x_{j-1} - x_*\|)/(1-q) \\ &\leq \|D_0\| + 2 \sum_{k=0}^{j-1} \gamma(\|x_k - x_0\|)/(1-q) \\ &\leq q - \tfrac{1}{2} + (1-q)/40 \leq \tfrac{1}{2}q \end{aligned}$$

which means that (2.19) holds also for $k = j$. Exactly the same derivation works for the Variable Metric Method since again with (2.19)

$$\|V(B_k, s_k, y_k, \phi_k) - V(B_k, s_k, s_k, \phi_k)\| \leq 3(1 + 8\|B_k\|)\gamma_k(1-q) \leq 40\ \gamma_k/(1-q)$$

and $\hat{\gamma}(\rho_0)$ is sufficiently small to absorb the larger factor. In either case we have the Banach lemma

$$\|B_j^{-1}\| \leq 1/(1 - \|D_j\|) \leq 1/(1 - q/2) < 2 \tag{2.22}$$

so that with (2.19), (2.8) and (2.21)

$$\begin{aligned} \|x_{j+1} - x_*\| &= \|x_j - x_* - B_j^{-1} g_j\| \\ &\leq \|B_j^{-1}\|\ \|(B_j - I)(x_j - x_*) + (x_j - x_* - g_j)\| \\ &\leq |\tfrac{1}{2}q + \gamma(\|x_j - x_*\|)|\,\|x_j - x_*\|/(1 - q/2) \leq q\|x_j - x_*\| . \end{aligned}$$

Thus we have shown that (2.17) and (2.19) hold also for $k = j$ and consequently for all $k \geq 0$. Finally we have by (2.10) and (2.22)

$$\begin{aligned} \overline{\lim}\ \|g_{k+1}\|/\|g_k\| &= \overline{\lim}\ \|y_k - s_k + g_k + s_k\|/\|B_k\ s_k\| \\ &\leq \overline{\lim}\ [\|y_k - s_k\| + \|(I - B_k)s_k\|]2/\|s_k\| \\ &\leq \overline{\lim}\ 4\gamma_k/(1-q) + \overline{\lim}\ 2\|D_k\ s_k\|/\|s_k\| \end{aligned}$$

which completes the proof as (2.19) and (2.20) have already been established. □

In stating Theorem 2.2 we have tried to make the conditions on B_0 as weak as possible. In general, i.e. if $B_0 \neq I$, the theorem applies with $D_0 = B_*^{-1} B_0 - I$ or $D_0 = B_*^{-\frac{1}{2}} B_0 B_*^{-\frac{1}{2}} - I$ in the nonsymmetric or optimization case respectively. Also the "Euclidean" norms in (2.17) and (2.18) must be replaced by topologically equivalent ellipsoidal norms depending on B_* . The linear convergence rate asserted in (2.17) is obviously rather pessimistic since it would be achieved even if B_k were kept constantly equal to B_0 . According to (2.18) the actual speed of convergence depends on the approximation errors for which we shall develop a much tighter bound in the next section.

3. The Asymptotic Rate of Convergence

The proof of superlinear convergence in the finite dimensional case (see [3] and [5]) is based on the observation that the Frobenius norm of the discrepancies $D_k = B_k - I$ is reduced at each step up to higher order terms. Moreover since this reduction is substantial whenever the approximation error $D_k s_k/\|s_k\|$ is significant, both tend to zero as k goes to infinity. In an infinite dimensional Hilbert space the Frobenius norm of D_k may be infinite and one can apply countably many low rank updates without ever obtaining a reasonable derivative approximation. However as suggested by the linear example in the introduction, things may not be so bad if D_0 is compact or has at least a comparatively small essential norm $\lambda_*(D_0)$ as defined by

$$\lambda_*(D) \equiv \inf \{\|D - C\| \mid \mathcal{B}(X,Y) \ni C \text{ compact}\} .$$

Because we have always $\operatorname{rank}(D_k - D_0)$ it follows from the conservation of the essential spectrum under compact perturbations that $\lambda_*(D_k)$ is the same for all k . Consequently the best bound on $\|D_k s_k\|/\|s_k\|$ we can possibly achieve without restricting the step directions $s_k/\|s_k\|$ is given by

$$\overline{\lim}\ \|D_k s_k\|/\|s_k\| \le \lambda_*(D_0) \le \|D_k\| .$$

As we shall see this inequality is in fact true for Broyden's method and a similar result holds in the optimization case.

In order to track the larger singular values of D_k we exploit the following variational characterization due to Weyl [18]. With S ranging over all subspaces of X , the j-th largest spectral value of a selfadjoint operator $A \in \mathcal{B}(X,X)$ is given by

$$\begin{aligned} \lambda_j(A) &= \inf_{\dim(S)<j} \ \sup_{0=v\in S^\perp} v^T A v / v^T v \\ &= \inf_{\dim(S)<j} \ \sup_{0\neq v\in S^\perp \cap s^\perp} v^T A v / v^T v \end{aligned} \tag{3.1}$$

where $s \in X$ may be any eigenvector such that

$$As = \mu s \quad \text{with} \quad \mu \le \lambda_*(A) .$$

In particular we find for the norm of the operators $A = D^T D$ and $\tilde{A} = \tilde{D}^T\tilde{D}$ with $\tilde{D} = D + U(B,s,s)$

$$\|\tilde{D}\|^2 = \lambda_1(\tilde{A}) = \sup_{0\neq v} v^T\tilde{A}v/v^Tv = \sup_{0\neq v\in s^\perp} (\|\tilde{D}v\|/\|v\|)^2$$

$$= \sup_{0\neq v\in s^\perp} (\|Dv\|/\|v\|)^2 \le \sup_{0\neq v} v^TAv/v^Tv = \|D\|^2$$

where we have used that $\tilde{D}s = 0 = \tilde{A}s$. Thus we have confirmed the relation (2.15). In the optimization case it is easier for study the eigenvalues of B and its positive definite inverse $H = B^{-1}$ rather than the singular values of $D = B - I$. In particular we find for

$$\tilde{B} \equiv B + V(B,s,s,\phi) \quad \text{with} \quad \tilde{B}s = s$$

that

$$\max\{1,\lambda_1(\tilde{B})\} \le \max\{1,\lambda_1(B)\} \tag{3.2}$$

since $\lambda_1(\tilde{B}) > 1$ implies by (3.1)

$$\lambda_1(\tilde{B}) = \sup_{0\neq v\in s^\perp} v^T\tilde{B}v/v^Tv$$

$$= \sup_{0\neq v\in s^\perp} [v^TBv - (1-\phi)(v^TBs)^2/s^TBs]/v^Tv$$

$$\le \sup_{0\neq v\in s^\perp} v^TBv/v^Tv \le \lambda_1(B) .$$

Moreover it is well known that for

$$\psi = (1-\phi)/[1-\phi(1 - s^THs\ s^TBs/(s^Ts)^2)]$$

$$\tilde{H} \equiv [B + V(B,s,s,\phi)]^{-1} = H + V(H,s,s,\psi)$$

so that due to the complete symmetry between B and H

$$\max\{1,\lambda_1(\tilde{H})\} \le \max\{1,\lambda_1(H)\} . \tag{3.3}$$

Since $1/\lambda_1(H) - 1$ and $\lambda_1(B) - 1$ respresent the smallest and largest eigenvalue of D it follows from (3.2) and (3.3) that $\|D + V(B,s,s,\phi)\| \le \|D\|$. This inequality confirms (2.16) and thus the local convergence result of the previous section. In order to determine the asymptotic convergence rate for both Broyden and Variable Metric Methods we make use of the following common framework.

Lemma 3.1. *Under the assumptions of Theorem 2.2 let the sequence* $\langle A_k\rangle_{k=0}^\infty \subset \mathcal{B}(X,X)$ *of positive semidefinite selfadjoint operators be defined by* $A_k = D_k^T D_k$ *in case of Broyden's method and* $A_k = B_k$ *or* $A_k = H_k = B_k^{-1}$ *in the case of the Variable Metric Method. Then there exist operators* $\langle \tilde{A}_k\rangle_{k=1}^\infty$ *of the same kind such that for a sequence* $\langle s_k\rangle_{k=1}^\infty \subset X - \{0\}$ *, some constant* c *and all* $k \ge 0$, $v \in X$

$$\operatorname{rank}(A_k - A_0) < \infty \tag{3.4}$$

$$\|A_{k+1} - \tilde{A}_{k+1}\| \le c\ \gamma_k \tag{3.5}$$

$$v^T\ \tilde{A}_{k+1}\ v \le v^T\ A_k\ v \quad \text{if} \quad v^T\ s_k = 0 \tag{3.6}$$

$$\tilde{A}_{k+1}\ s_k = \mu\ s_k \qquad \text{for constant } \mu \tag{3.7}$$

where $\mu = 0$ *and* $\mu = 1$ *in the Broyden and Variable Metric case respectively.*

<u>Proof</u>. In the Broyden case we may set

$$\tilde{A}_{k+1} \equiv \tilde{D}_{k+1}^T\ \tilde{D}_{k+1} \quad \text{with} \quad \tilde{D}_{k+1} = D_k + U(B_k, s_k, s_k)$$

so that by Lemma 2.1

$$\|D_{k+1} - \tilde{D}_{k+1}\| = \|U(B_k, s_k, y_k) - U^*(B_k, s_k, s_k)\| \le 2\ \gamma_k/(1-q)$$

and consequently

$$\|A_{k+1} - \tilde{A}_{k+1}\| \le \|D_{k+1} - \tilde{D}_{k+1}\|\|D_{k+1} + \tilde{D}_{k+1}\| \le c\ \gamma_k$$

where $c \equiv 4[\sup\|B_k\| + 1 + \gamma_0/(1-q)]/(1-q)$. The remaining two properties hold since by definition of U

$$v^T\tilde{A}_{k+1}v = \|\tilde{D}_{k+1}v\|^2 = \|D_k v\|^2 = v^T A_k v \quad \text{if} \quad v^T s_k = 0$$

and obviously $\tilde{A}_{k+1}\ s_k = \tilde{D}_{k+1}^T\ \tilde{D}_{k+1}\ s_k = \tilde{D}_{k+1}^T\ 0 = 0$. To establish these relations in the Variable Metric Case for $A_k = B_k$ we set $\tilde{A}_{k+1} = B_k + V(B_k, s_k, s_k, \phi_k)$ so that by Lemma 2.1

$$\|A_{k+1} - \tilde{A}_{k+1}\| = \|V(B_k, s_k, y_k, \phi_k) - V(B_k, s_k, s_k, \phi_k)\| \le c\ \gamma_k \tag{3.8}$$

where $c \equiv 3(1 + 8\sup\|B_k\|)/(1-q) \le 40/(1-q)$. By inspection of (2.7) we derive from $v^T s_k = 0$ that

$$v^T\tilde{A}_{k+1}v = v^T A_k v - (1-\phi)(v^T A_k s)^2/s_k^T A_k s_k \le v^T A_k v$$

and obviously $\tilde{A}_{k+1}\ s_k = s_k$. Finally in the case $A_k = H_k$ we set

$$\tilde{A}_{k+1} \equiv [B_k + V(B_k, s_k, s_k, \phi_k)]^{-1} = H_k + V(H_k, s_k, s_k, \psi_k)$$

where $\psi_k = (1-\phi_k)/[1 - \phi_k(1 - s_k^T H_k s_k\ s_k^T B_k s_k/(s_k^T s_k)^2)]$. Now it can be easily checked that

$$\sup\|\tilde{A}_k\| \le 1 + 3\sup\|H_k\| < \infty .$$

Thus we have with $A_{k+1}^{-1} = B_{k+1}$

$$\begin{aligned}\|A_{k+1} - \tilde{A}_{k+1}\| &= \|A_{k+1}[V(B_k, s_k, y_k, \phi_k) - V(B_k, s_k, s_k, \phi_k)]\tilde{A}_{k+1}\| \\ &\le c\ \sup\|\tilde{A}_k\|\|A_k\|\ \gamma_k\end{aligned}$$

where c is the constant occurring in (3.8). Again due to the symmetry between B_k and H_k the last two assertions follow exactly as in the case $A_k = B_k$. □

Because of $\operatorname{rank}(A_k - A_0) < \infty$ the essential norm $\lambda_*(A_k)$ is the same for all $k \geq 0$. In order to conveniently analyse the larger singular values of the A_k we need the additional assumption that $\lambda_*(A_0) \geq \mu$. Since A_0 is required to be semi-definite this condition is automatically ratified in the Broyden case where $\mu = 0$. Now suppose that in the optimization case $\lambda_*(B_0) < 1$ or $\lambda_*(H_0) < 1$. Then we can extend the original equation $g(x) = 0$ to the equivalent system

$$\hat{g}(x,z) \equiv (g(x),z) = 0 \quad \text{with} \quad z \in \ell^2$$

which has obviously the same regularity and smoothness properties at the root $\hat{x}_* = (x_*,0) \in X \times \ell^2$. Moreover the Variable Metric Method generates on $\hat{g} = 0$ the corresponding iterates $(x_k,0)$ if started from

$$\hat{x}_0 = (x_0,0) \quad \text{with} \quad \hat{B}_0 \equiv B_0 \times I .$$

Since $\lambda_*(\hat{B}_0) = \max\{1,\lambda_*(B_0)\}$ and for $\hat{H}_0 = \hat{B}_0^{-1}, \lambda_*(\hat{H}_0) = \max\{1,\lambda_*(H_0)\}$ the condition $\lambda_*(A_0) \geq \mu = 1$ is now met for both choices $A_k = \hat{B}_k$ and $A_k = \hat{H}_k$. In particular this embedding allows us to apply our result to finite-dimensional problems where necessarily $\lambda_*(B_0) = 0 = \lambda_*(H_0)$ like for any other compact operator. After this theoretical extension we shall how assume without loss of generality that $\lambda_*(A_0) \geq \mu$.

The definition (3.1) provides us always with a countable chain of repeated spectral values

$$\lambda_1(A_k) \geq \lambda_2(A_k) \geq \quad \geq \lambda_j(A_k) \geq \quad \geq \lambda_* = \lim \lambda_j(A_k) \geq \mu \geq 0 .$$

Due to their isolation the $\lambda_j(A_k) > \lambda_*$ are proper eigenvalues in which case the infinum and supremum in (3.1) are attained as minimum and maximum respectively. However there may be only finitely many such $\lambda_j(A_k) > \lambda_*$ and the eigenspace of A_k associated with λ_* may not allow us to assign an eigenvector to each $\lambda_j(A_k) = \lambda_*$. By an extension similar to the one described above this difficulty can be easily circumvented as is done in the proof of the following main result.

<u>Theorem 3.2</u>. *Let* $\langle A_k \rangle_{k=0}^{\infty} \subset B(X,X)$ *be any sequence of positive semi-definite selfadjoint operators such that the conditions* (3.4)-(3.7) *hold with*

$$\lambda_*(A_0) \geq \mu \quad \text{and} \quad \sum_{k=0}^{\infty} \lambda_k < \infty . \tag{3.9}$$

Then the spectral values $\lambda_j(A_k)$ *have the bounded deterioration property*

$$\lambda_j(A_{k+1}) \le \lambda_j(A_k) + c\,\gamma_k \tag{3.10}$$

and there exist limits

$$\lambda_j^* \equiv \lim_{k\to\infty} \lambda_j(A_k) \ge \lambda_{j+1}^* . \tag{3.11}$$

Moreover we have

$$\lambda_* = \lambda_*^* = \lim_{j\to\infty} \lambda_j^* \tag{3.12}$$

and as a consequence

$$\overline{\lim}\ s_k^T A_k s_k / s_k^T s_k \le \lambda_* = \lambda_*(A_0) . \tag{3.13}$$

Proof. For the sake of simplicity we would like each $\lambda_j(A_k)$ to be associated with an eigenvector that is orthogonal to the corresponding eigenvectors of all previous $\lambda_i(A_k)$ with $i<j$. This is always possible unless for some (and consequently all) k

$$\lambda_* \in \{\lambda_j(A_k)\}_{j=1}^{\infty} \quad \text{and} \quad \text{nullity}(A_k - \lambda_* I) < \infty$$

which can only happen if $\lambda_* > 0$. In this exceptional case we may replace A_k by

$$\hat{A}_k \equiv A_k \times \lambda_* I : X \times \ell^2 \to Y \times \ell^2 ,$$

modify $\tilde{A}_{k+1}$ accordingly and identify the s_k with $(s_k, 0) \in X \times \ell^2$. Then we have

$$\lambda_j(\hat{A}_k) = \lambda_j(A_k) \quad \text{for all} \quad j,k$$

and each $\hat{A}_k$ has an eigenspace containing $\{0\} \times \ell^2$ associated with $\lambda_*(\hat{A}_k) = \lambda_*(A_0)$. Moreover it can be easily checked that the conditions (3.4)-(3.7) and (3.9) as well as the assertion (3.13) apply equivalently to the original and extended system. Therefore we can assume without any loss of generality that all infima and suprema in (3.1) are attained as minima and maxima respectively.

Combining (3.1) and (3.6) we find that

$$\begin{aligned} \lambda_j(\tilde{A}_{k+1}) &= \min_{\dim(S)<j} \ \max_{0\ne v\in S^\perp \cap s_k^\perp} v^T \tilde{A}_{k+1} v / v^T v \\ &\le \min_{\dim(S)<j} \ \max_{0\ne v\in S^\perp \cap s_k^\perp} v^T A_k v / v^T v \le \lambda_j(A_k) \end{aligned} \tag{3.14}$$

which implies (3.10) by (3.5) since clearly for any $S \subset X$

$$\max_{0\ne v\in S^\perp} v^T A_{k+1} v / v^T v \le \|A_{k+1} - \tilde{A}_{k+1}\| + \max_{0\ne v\in S^\perp} v^T \tilde{A}_{k+1} v / v^T v .$$

The existence of the limits λ_j^* follows by Lemma 3.3 in [5]. Since the sequence $\langle \lambda_j^* \rangle$ is nonincreasing and bounded below by $\lambda_* = \lambda_*(A_k)$ it must reach a limit $\tilde{\lambda}_* \ge \lambda_*$. Now if $\tilde{\lambda}_*$ were greater than λ_*

there would be an index m such that

$$c \sum_{k=m}^{\infty} \gamma_k \le \tfrac{1}{3}(\tilde{\lambda}_* - \lambda_*) \ .$$

Since $\lambda_j(A_m) \to \lambda_*(A_m) = \lambda_*$ there exists an index j such that

$$\lambda_j(A_m) \le \lambda_* + \tfrac{1}{3}(\tilde{\lambda}_* - \lambda_*) \ .$$

Substituting these two inequalities into (3.9) we obtain for all $k > m$

$$\lambda_j(A_k) \le \lambda_j(A_m) + c \sum_{i=m}^{k-1} \gamma_i \le \frac{1}{3}\lambda_* + \frac{2}{3}\tilde{\lambda}_* < \tilde{\lambda}_*$$

which contradicts the assumption $\lambda_j^* \ge \tilde{\lambda}_* > \lambda_*$. In order to establish the main assertion (3.13) we use the partial traces

$$\tau_m(A) \equiv \sum_{j=1}^{m} \lambda_j(A) = \max \{ \sum_{j=1}^{m} v_j^T A v_j \,|\, v_i^T v_j = \delta_{ij} \} \qquad (3.15)$$

with the limits

$$\tau_m^* \equiv \lim_{k\to\infty} \tau_m(A_k) = \sum_{j=1}^{m} \lambda_j^* \ge m\lambda_*(A_0) \ .$$

Now let us choose an eigenvector $\{v_j\}_{j=1}^m$ that corresponds to the largest m eigenvalues of $\tilde{A}_{k+1}$ and form together with $s_k/\|s_k\|$ an orthonormal family. Thus it follows by (3.14) and (3.15) that

$$\tau_{m+1}(A_k) \ge s_k^T A_k s_k / s_k^T s_k + \sum_{j=1}^{m} v_j^T A_k v_j$$

$$\ge s_k^T A_k s_k / s_k^T s_k + \tau_m(\tilde{A}_{k+1}) \ .$$

Since by (3.5)

$$|\tau_m(A_{k+1}) - \tau_m(\tilde{A}_{k+1})| \le m\, c\, \gamma_k$$

we have the limit

$$\limsup_{k\to\infty} s_k^T A_k s_k / s_k^T s_k \le \lim_{m\to\infty}[\lim_{k\to\infty} \tau_{m+1}(A_k) - \tau_m(A_{k+1})]$$

$$\equiv \lim_{m\to\infty}(\tau_{m+1}^* - \tau_m^*) = \lim_{m\to\infty} \lambda_{m+1}^* = \lambda_*(A_0)$$

where the last equation follows from (3.12). □

The bounded deterioration property (3.10) was originally introduced by Dennis and Moré for the Frobenius norm of the discrepancies D_k. If g is linear so that $\gamma_k = 0$, all eigenvalues $\lambda_j(A_k)$ are reduced or unchanged at each step. This observation appears to be new for Broyden's method where the roots $\sqrt{\lambda_j(A_k)}$ represent the larger singular values of the nonsymmetric operators D_k. In the optimization case, where g is the gradient of a quadratic objective function, our analysis generalizes Fletcher's result [8] that Variable Metric Updates have "Property 1". The larger eigenvalues $\lambda_j(B_k) \ge 1$ of B_k

are nonincreasing and the smaller eigenvalues $1/\lambda_j(H_k) \le 1$ of B_k are nondecreasing as shown below.

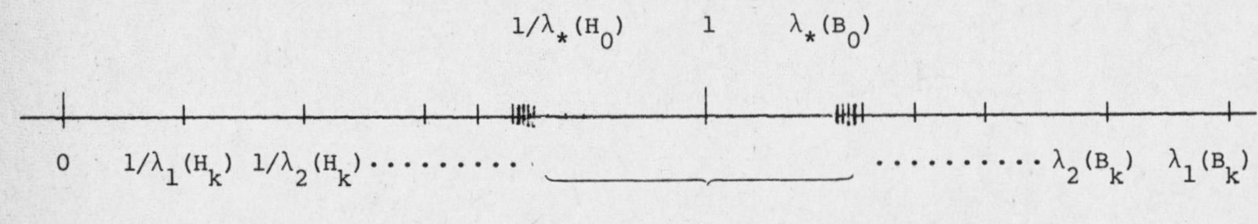

According to (3.13) secant updating effectively eliminates all eigenvalues above $\lambda_*(A_0)$ which leads to the following Corollary.

Corollary 3.3. *Under the assumptions of Theorem 2.2 it follows from Theorem 3.2 that for Broyden's method*

$$\overline{\lim}\ \|g_{k+1}\|/\|g_k\| \le 2\ \lambda_*(D_0) \tag{3.16}$$

and for the Variable Metric Method

$$\overline{\lim}\ \|g_{k+1}\|/\|g_k\| \le \frac{5}{2}\sqrt{\lambda_*(B_0) + \lambda_*(B_0^{-1}) - 2} \le 5\sqrt{\lambda_*(D_0)} \tag{3.17}$$

where in both cases $D_0 = B_0 - I$ *and*

$$\lambda_*(D_0) = \inf\{\|D_0 - C\| \mid \mathcal{B}(X,Y) \ni C \text{ compact}\} .$$

Moreover, if D_0 *is compact, i.e.* $\lambda_*(D_0) = 0$ *, we have Q-superlinear convergence in that*

$$\lim_{k\to\infty} \|g_{k+1}\|/\|g_k\| = 0 = \lim_{k\to\infty} \|x_{k+1} - x_*\|/\|x_k - x_*\| . \tag{3.18}$$

Proof. In the Broyden case we have

$$\|D_k s_k\|/\|s_k\| = (s_j^T D_k^T D_k s_k / s_k^T s_k)^{\frac{1}{2}} = (s_k^T A_k s_k / s_k^T s_k)^{\frac{1}{2}}$$

so that (3.16) follows immediately from (2.18) and (3.13). In the symmetric case we have with (2.19)

$$\|D_k s_k\|^2 = \|(B_k - I)s_k\|^2 = \|B_k^{\frac{1}{2}}(B_k^{\frac{1}{2}} - H_k^{\frac{1}{2}})s_k\|^2$$
$$\le \|B_k\|\|(B_k^{\frac{1}{2}} - H_k^{\frac{1}{2}})s_k\|^2 \le \frac{3}{2}(s_k^T B_k s_k - 2 s_k^T s_k + s_k^T H_k s_k)$$

so that again by (3.13) for $A_k = B_k$ and $A_k = H_k$

$$\overline{\lim}\ (\|D_k s_k\|/\|s_k\|)^2 \le \frac{3}{2}(\overline{\lim}\ s_k^T B_k s_k / s_k^T s_k + \overline{\lim}\ s_k^T H_k s_k / s_k^T s_k - 2)$$
$$\le \frac{3}{2}(\lambda_*(B_0) + \lambda_*(H_0) - 2) .$$

This implies the first inequality (3.17) by (2.18) since $2\sqrt{3/2} < 5/2$. By comparing the spectra of B_0 and H_0 with the spectrum of D_0 we find that

$$\lambda_*(B_0) \le 1 + \lambda_*(D_0) \quad \text{and} \quad \lambda_*(H_0) \le 1/(1 - \lambda_*(D_0)) .$$

Thus we obtain with $\lambda_*(D_0) \equiv \|D_0\| \le \frac{1}{2}$ that

$$\lambda_*(B_0) + \lambda_*(H_0) - 2 \le \lambda_*(D_0) - 1 + 1/(1 - \lambda_*(D_0))$$
$$\le \lambda_*(D_0)(2 - \lambda_*(D_0))/(1 - \lambda_*(D_0)) \le 3\lambda_*(D_0)$$

which implies the second inequality in (3.17) as $2 \cdot \sqrt{3 \cdot 3/2} = 3 \cdot \sqrt{2} < 5$. Finally we note that the first equation in (3.18) follows immediately from (3.16), (3.17) and the fact that by (2.8) in any case $\lim \|g_k\|/\|x_k - x_*\| = 1$. □

In the general case where $B_* \neq I$ the assertions (3.16) and (3.17) hold with $D_0 = B_*^{-1}B_0 - I$ or $D_0 = B_*^{-\frac{1}{2}}B_0B_*^{-\frac{1}{2}} - I$ and either the norms on g_k or the constants on the right hand sides adjusted accordingly. However the superlinear convergence result (3.18) like the compactness assumption on D_0 apply equivalently for all topologically equivalent norms. It was shown in [10] under stronger differentiability assumptions on g that in the case of Broyden's method $\|g_k\| \le (c/k)^{k/p}$ for some c, whenever the singular values $\sqrt{\lambda_j(A_0)}$ of D_0 are $p \ge 2$ summable. It seems quite likely that a similar R-superlinear convergence result can be obtained for Variable Metric Methods.

Summary and Conclusion

It was observed in the introduction that the speed of convergence, achieved by Newton's method on discretized problems, may be determined by the asymptotic rate of its conceptual analog on the underlying operator equation $g(x) = 0$. Another example showed that in a Hilbert space setting the usual invertibility and smoothness assumption on the Frechet derivative $g'(x)$ are no longer sufficient for the local and Q-superlinear convergence of classical secant methods. However, using bounded deterioration with respect to the induced operator norm, local and Q-linear convergence could be established for full step iterations based on Broyden's update or Variable Metric formulae. Besides requiring bounded invertibility of the derivative $B_* = g'(x_*)$ at the root $x_* \in g^{-1}(0)$ we had to assume that the initial approximation B_0 was chosen such that the discrepancy

$$D_0 = B_*^{-1}B_0 - I \quad \text{or} \quad D_0 = B_*^{-\frac{1}{2}}B_0B_*^{-\frac{1}{2}} - I$$

is less than $\frac{1}{2}$ in the induced operator norm. The asymptotic rate of convergence was found to depend on the essential norm

$$\lambda_*(D_0) = \inf\{\|D_0 - C\| \mid \mathcal{B}(X,Y) \ni C \text{ compact}\}$$

which means that secant updating effectively eliminates all singular values greater than $\lambda_*(D_0)$. In particular we obtain Q-superlinear convergence if D_0 is compact which can be ensured in the case of mildly nonlinear problems such as integral equations of the second kind. On fully nonlinear problems such as implicit differential equations, the low rank updates are unable to keep up with noncompact changes in the derivative. On the corresponding discretizations the superlinear rate of convergence will only eventuate at the truncation error level. However better results might be achievable by sparse or partitioned updating techniques like those proposed in [15] and [9].

Acknowledgement: The author had the benefit of several discussions with Chris Beattie, VPI & SU, regarding the variational characterization of eigenvalues.

References

[1] E. Allgower and K. Böhmer, "A mesh independence principle for operator equations and their discretizations", Preprint, Department of Mathematics, Colorado State University (1984).

[2] C.G. Broyden, "A class of methods for solving nonlinear simultaneous equations", *Mathematics of Computation*, Vol. 19 (1965), pp.577-593.

[3] C.G. Broyden, J.E. Dennis and J.J. Moré, "On the local and superlinear convergence of quasi-Newton methods", *J. Inst. Math. Appl.*, Vol. 12 (1973), pp.223-245.

[4] J.W. Daniel, "The conjugate gradient method for linear and nonlinear operator equations", *SINUM*, Vol. 4 (1967), pp.10-26.

[5] J.E. Dennis and J.J. Moré, "A characterization of superlinear convergence and its application to quasi-Newton methods", *Mathematics of Computation*, Vol. 28 (1974), pp.543-560.

[6] J.E. Dennis and J.J. Moré, "Quasi-Newton methods. Motivation and theory", *SIAM Review*, Vol. 19 (1977), pp.46-89.

[7] J.E. Dennis and R.B. Schnabel, *Numerical Methods for Unconstrained Optimization and Nonlinear Equations*, Prentice-Hall, Englewood Cliffs, Prentice-Hall Series in Computational Mathematics, 1983.

[8] R. Fletcher, "A new approach to Variable Metric Algorithms", *Comp. J.*, Vol. 13 (1970), pp.317-322.

[9] A. Griewank and Ph. Toint, "Local convergence analysis for partitioned quasi-Newton updates", *Numerische Mathematik*, Vol. 39 (1982), pp.429-448.

[10] A. Griewank, "The local convergence of Broyden's Method on Lipschitzian problems in Hilbert spaces", to appear in *SINUM*.

[11] M. Hayes, "Iterative methods for solving nonlinear problems in

Hilbert space", in Contribution to the Solution of Linear Systems and the Determination of Eigenvalues (O. Taussky, ed.), *Appl. Math. Series* 39, National Bureau of Standards, Washington DC, 1954.

[12] M.J.D. Powell, "On the rate of convergence of variable metric algorithms for unconstrained minimization", Technical Report DAMTP 1983/NAF (1983).

[13] Ge Ren-Pu and M.J.D. Powell, "The convergence of variable metric matrices in unconstrained optimization", *Math. Programming*, Vol. 27 (1983), pp.233-243.

[14] E. Sachs, "Broyden's method in Hilbert spaces", Preprint, Mathematics Department, North Carolina State, Raleigh, N.C. (1984).

[15] L.K. Schubert, "Modification of a quasi-Newton method for nonlinear equations with a sparse Jacobian", *Math. of Comp.*, Vol. 24 (1970), pp.27-30.

[16] J. Stoer, "Two examples on the convergence of certain rank-2 minimization methods for quadratic functionals in Hilbert space", *Linear Algebra and Its Applications*, Vol. 28 (1979), pp.217-222.

[17] J. Stoer, "The convergence of matrices generated by rank-2 methods from the restricted B-class of Broyden", *Numerische Mathematik*, Vol. 44 (1984), pp.37-52.

[18] A. Weinstein and W. Stenger, *Intermediate Problems for Eigenvalues Theory and Ramifications*, Academic Press, New York (1972).

[19] R. Winther, "Some superlinear convergence results for the conjugate gradient method", *SINUM*, Vol. 17 (1980), pp.14-18.

[20] R. Winther, "A numerical Galerkin method for a parabolic problem", Ph.D Dissertation, Cornell University, New York, 1977

[21] K. Yosida, *Functional Analysis*, Grundlehren der mathematischen Wissenschaften 123, Springer-Verlag, Berlin, Heidelberg, New York, 1980.

THE CONSTRUCTION OF PRECONDITIONERS FOR ELLIPTIC PROBLEMS BY SUBSTRUCTURING

by

J.H. Bramble
Cornell University

In this talk I will consider as a model problem the Dirichlet problem for a second order uniformly elliptic equation in two dimensions. Let Ω be a bounded domain in R^2 which, for the sake of exposition, has a polygonal boundary $\partial\Omega$. Thus we shall consider the problem

$$1)\qquad \begin{aligned} Lu &= f \quad \text{in} \quad \Omega \\ u &= 0 \quad \text{on} \quad \partial\Omega \end{aligned}$$

where

$$Lv = -\sum_{i,j=1}^{2} \frac{\partial}{\partial x_i}\left(a_{ij}\frac{\partial v}{\partial x_j}\right)$$

with a_{ij} uniformly positive definite, bounded and piecewise smooth on Ω. The generalized Dirichlet form is given by

$$A(v,\phi) = \sum_{i,j=1}^{2} \int_\Omega a_{ij} \frac{\partial v}{\partial x_i} \frac{\partial \phi}{\partial x_j}\, dx$$

defined for all v and ϕ in the Sobolev space $H^1(\Omega)$ (the space of distributions with square integrable first derivatives) The $L^2(\Omega)$ inner product is given by

$$(v,\phi) = \int_\Omega v\phi dx \, .$$

The subspace $H_0^1(\Omega)$ of $H^1(\Omega)$ is the completion of the smooth functions with support in Ω with respect to the norm in $H^1(\Omega)$. By integration by parts the problem defined by 1) may be written in weak form: Find $u \varepsilon H_0^1(\Omega)$ such that

$$2)\qquad A(u,\phi) = (f,\phi)$$

for all $\phi \varepsilon H_0^1(\Omega)$. This leads immediately to the standard Galerkin approximation. Let S_h be a finite dimensional subspace of $H_0^1(\Omega)$. The Galerkin approximation is defined as the solution of the following problem: find $U \varepsilon S_h$ such that

$$3)\qquad A(U,\chi) = (f,\chi)$$

for all $\chi \varepsilon S_h$. Once a basis $\{\chi_i\}_{i=1}^N$ for S_h is chosen, 3) leads to a system of linear algebraic equations. Write

$$U = \sum_{i=1}^{N} \alpha_i \chi_i \quad .$$

Then 3) becomes

$$4) \qquad \sum_{i=1}^{N} \alpha_i A(\chi_i, \chi_j) = (f, \chi_j)$$

$j=1,\ldots,N$ which is a linear system for the determination of the coefficients α_i, $i=1,\ldots,N$.

It is well known that for a wide class of approximation spaces, S_h, U will be a good approximation to u. We shall consider certain spaces S_h for which we may also develop efficient algorithms for the solution of the underlying linear system 4). Such a strategy is quite usual. For example, for the Laplace operator on a rectangular region, a subspace S_h of piecewise linear functions on a uniform triangulation leads to the usual 5-point approximation to the Laplacian. The resulting equations may be solved "fast" using, for example, fast Fourier transform techniques. Other choices of S_h in this case may lead to good approximate solutions, but they are perhaps not obtained so efficiently. Another example of a special choice of S_h which leads often to a fast algorithm is one which may be thought of as connected with a nested set of grids. For such spaces, S_h, a "multigrid" algorithm may be applied.

I shall discuss another technique which has some features of its own. The underlying method which I will consider is a preconditioned iterative method. The choice of particular iterative method within a certain class is not essential, but for the purpose of this talk we may think of the well-known conjugate gradient method which is often used in practice. Roughly, this may be described as follow. Let A be the $N \times N$ matrix with entries $A(\chi_i, \chi_j)$, $\alpha = (\alpha_1, \ldots, \alpha_N)$ and F the vector with components (f, χ_j). Then 4) may be written

$$5) \qquad A\alpha = F \; .$$

Generally, the matrix A is not well conditioned so that a direct application of the conjugate gradient method to the symmetric positive definite system 5) will not be a very efficient algorithm. The preconditioned conjugate gradient method (PCG) consists of choosing a positive definite symmetric matrix B and writing 5) as

$$6) \qquad B^{-1}A\alpha = B^{-1}F \; .$$

In the context of this talk the matrix B will be associated with another bilinear form $B(\cdot,\cdot)$ defined on $S_h \times S_h$. The system 6) is symmetric with respect to the inner product defined by

$$7) \qquad [\alpha,\beta] \equiv \sum_{i,j=1}^{N} B_{ij}\alpha_i\beta_j \quad .$$

Thus the conjugate gradient method may be applied to 6) with respect to 7). The importance of making a "good" choice for B is well known. The matrix B should have two main properties. First, the solution of the problem

$$8) \qquad B\beta = b$$

should be easy to obtain. This is tantamount to applying the operator B^{-1} to obtain the vector b. Secondly, B should be spectrally close to A in the following sense. With λ_0 and λ_1 defined so that

$$\lambda_0[\beta,\beta] \leqslant [B^{-1}A\beta,\beta] \leqslant \lambda_1 [\beta,\beta]$$

the "condition number" $K=\frac{\lambda_1}{\lambda_0}$ should grow at most "slowly" with N. In terms of the form $B(\cdot,\cdot)$ the first property means that the solution W of

$$B(W,\chi) = (g,\chi) \quad , \quad \forall\ \chi\varepsilon S_h$$

for a given function g should be easier to obtain than the solution of 2). The spectral condition, in terms of the forms, is

$$\lambda_0 B(V,V) \leqslant A(V,V) \leqslant \lambda_1 B(V,V)$$

for all V S_h.

These two properties will guarantee, firstly, that the work per step in applying the conjugate gradient method (as an iterative method) will be small, and, secondly, that the number of steps to reduce the error to a given size will be also small so that an efficient algorithm will result.

The principal point which I will discuss is joint work with J. Pasciak and A. Schatz in which we focus our attention on techniques for the construction of preconditioners B possessing qualities described above.

We shall suppose that the domain Ω has been triangulated and that the subspace S_h consists of continuous, piecewise linear functions. Further, we assume that this triangulation is such that the region Ω

may be written as the union of regions Ω_j with maximum diameter d which are either triangles or rectangles whose sides coincide with the mesh lines in the original triangulation. The vertices of the Ω_j will be labeled v_k (ordered in some way) and the segments connecting the vertices will be called Γ_{ij} with endpoints v_i and v_j taken, of course, only when Γ_{ij} is an edge of some Ω_k. The following figure should help to clarify these definitions.

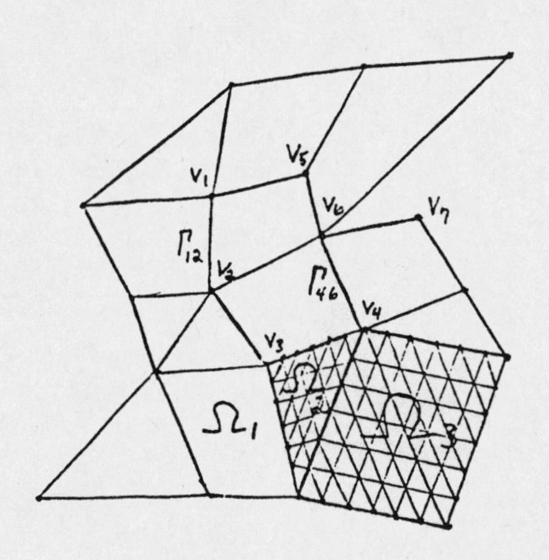

We assume that each Ω_k has a triangulation inherited from the original triangulation and we denote by $S_h^0(\Omega_k)$ the subspace of S_h consisting of elements of S_h which vanish outside of Ω_k and in particular on $\partial\Omega_k$. We construct our preconditioner B by constructing its corresponding bilinear form $B(\cdot,\cdot)$ defined on $S_h \times S_h$. To show how we do this we decompose an arbitrary $V \varepsilon S_h$ as follows: Write $V = V_P + V_H$ where, $V_P = 0$ on $\partial\Omega_k$ and satisfies

$$A_k(V_P,\chi) = A_k(V,\chi) \quad , \quad \forall\ \chi \varepsilon S_h^0(\Omega_k)$$

with

$$A_k(V_P,\chi) \equiv \sum_{i,j=1}^{2} \int_\Omega a_{ij}^k \frac{\partial V_P}{\partial x_i} \frac{\partial \chi}{\partial x_j} \, dx \ .$$

Here a_{ij}^k is a constant positive definite matrix for each k. Notice that V_P is determined on Ω_k by the values of V on Ω_k and that

$$A_k(V_H,\chi) = 0 \quad , \quad \forall\ \chi \varepsilon S_h^0(\Omega_k) \quad .$$

Thus, on each Ω_k, V is decomposed into a function V_P which vanishes on $\partial\Omega_k$ and a function V_H which satisfies the above homogeneous equations. With a slight abuse of terminology we shall refer to such a function as "discrete harmonic".

We next decompose V_H on Ω_k into $V_H=V_e+V_v$, where V_v is the discrete harmonic function whose values on $\partial\Omega_k$ are the linear function along each Γ_{ij} with the same values as V at the vertices. Thus V_e is a discrete harmonic function in Ω_k for each k which vanishes at all of the vertices. Hence, once V_e and V_v are determined on all of the Γ_{ij}'s, $V_H=V_e+V_v$ is then determined in each Ω_k by solving a Dirichlet problem only on Ω_k with respect to a constant coefficient operator.

Before defining the form $B(\cdot,\cdot)$ we note that for any discrete harmonic function W on Ω_k

$$\alpha_0 A_k(W,W) \leqslant |W|^2_{1/2,\partial\Omega_k} \leqslant \alpha_1 A_k(W,W) \tag{9}$$

where α_0 and α_1 are positive constants and $|\cdot|_{1/2,\partial\Omega_k}$ is the norm on the Sobolev space $H^{1/2}(\partial\Omega_k)$. Now it may be shown that if W=0 at the vertices then the norm $|W|^2_{1/2,\partial\Omega_k}$ may be replaced in 9) by $\sum_{\Gamma_{ij}\subset\partial\Omega_k} \langle \ell_0^{1/2}W,W\rangle_{\Gamma_{ij}}$ with new values of α_0 and α_1 such that $\alpha_1/\alpha_0=((\ell n\,\frac{d}{h})^2)$. Here ℓ_0 is the operator defined on the restriction of functions in S_h to the $\cup\Gamma_{ij}$, which vanish at the vertices. The operator is given, for each Γ_{ij} by

$$\langle \ell_0 W,\chi\rangle_{\Gamma_{ij}} = \langle W',\chi'\rangle_{\Gamma_{ij}}$$

for all appropriate $\chi\varepsilon S_h$ vanishing at the vertices and the prime denotes differentiation along Γ_{ij}. Also

$$\langle \phi, \psi\rangle_{\Gamma_{ij}} = \int_{\Gamma_{ij}} \phi\psi\, dx \quad .$$

A key point here is that $\ell_0^{-1/2}$ on each Γ_{ij} may be computed by means of the fast Fourier transform.

Finally, on Ω_k, $A_k(V_v,V_v)$ is bounded above and below by constants times the expression

$$\sum_{\Gamma_{ij}\subset\partial\Omega_k} (V_v(V_i)-V_v(V_j))^2 \quad .$$

With the above statements in mind we now define the form $B(\cdot,\cdot)$ as follows:

$$B(V,\phi)=\widetilde{A}(V_P,\phi_P)+\sum_{\Gamma_{ij}} \langle \ell_0^{1/2} V_e,\phi_e \rangle_{\Gamma_{ij}} + \sum_{\Gamma_{ij}} (V_v(V_i)-V_v(V_j))\,(\phi_v(V_i)-\phi_v(V_j))$$

and $\widetilde{A}(\cdot,\cdot)=A_k(\cdot,\cdot)$ on the restriction of S_h to Ω_k.

We can prove that this bilinear form has the following properties

$$\lambda_0 B(V,V) \leqslant A(V,V) \leqslant \lambda_1 B(V,V) \quad ,$$

where $K=\lambda_1/\lambda_0=O((\ell n\,\frac{d}{h})^3)$. Thus the condition number grows at most like $(\ell n\,\frac{d}{h})^3$ as h tends to zero. This corresponds to the second of the two desirable properties mentioned earlier.

The property corresponding to the first property previously discussed is that the problem

$$10) \qquad B(V,\phi) = (g,\phi) \quad , \quad \forall\ \phi\varepsilon S_h$$

is much more easily solved than is the original. This means that the solution of the corresponding matrix equation 8) is relatively easy to obtain.

In order to see how to solve 10) efficiently, we shall see that the defining equations have been chosen to conveniently lend themselves to a "block Gauss elimination" procedure. This is most easily understood by describing the process used to solve 10).

As we can see, we want to find V_P and V_H. The function V_P is obtained on each Ω_k independently of V_H by choosing basis functions in $S_h^0(\Omega_k)$ and solving the corresponding constant coefficient subproblems, which themselves are independent of each other.

With V_P now known, we are left with the equation

$$11) \qquad \sum_{\Gamma_{ij}} \langle \ell_0^{1/2} V_e,\phi_e \rangle_{\Gamma_{ij}} + \sum_{\Gamma_{ij}} (V_v(V_i)-V_v(V_j))\,(\phi_v(V_i)-\phi_v(V_j))$$

$$= (g,\phi) - \widetilde{A}(V_P,\phi_P)$$

$$= (g,\phi) - \widetilde{A}(V_P,\phi)$$

the last equality holding since $\widetilde{A}(V_P,\phi_H)=0$. Notice that the value of $(g,\phi)-\widetilde{A}(V_p,\phi)$, for each ϕ, depends only on the value of ϕ on the Γ_{ij}'s. Thus 11) gives rise to a set of equations on the restriction of S_h to $\cup\Gamma_{ij}$. To solve these equations, we proceed as follows: For each Γ_{ij} choose a subspace of S_h whose elements vanish on all other Γ's and in particular at the endpoints of Γ_{ij}. Thus, on this subspace, 11) reduces to

$$< \ell_0^{1/2} V_e,\phi >_{\Gamma_{ij}} = (g,\phi)-\widetilde{A}(V_P,\phi)$$

for each Γ_{ij}. This equation is easily solved for V_e on each Γ_{ij} by means of the fast Fourier transform.

Finally, we choose a subspace of S_h which reduces to linear functions between the endpoints of each Γ_{ij}. Clearly such a subspace has dimension equal to the number of interior nodes on Γ_{ij}. For each ϕ in this subspace, $\phi_e=0$ and 11) reduces to

$$\sum_{\Gamma_{ij}} (V_v(V_i)-V_v(V_j))\,(\phi_v(V_i)-\phi_v(V_j)) = (g,\phi)-\widetilde{A}(V_P,\phi) .$$

A basis for this subspace may be chosen as follows: Choose $\{\phi^1,\ldots,\phi^M\}$, where M is the number of vertices not on $\partial\Omega$ and $\phi^i(V_j)=\delta_{ij}$ where $\delta_{ij}=1$ if $i=j$ and 0 if $i\neq j$. This choice gives rise to a set of difference equation on the vertex "mesh points" which may be solved for the values V_v there. The values V_v at the vertices determine V_v on the edges and hence $V_H=V_e+V_v$ is known on all of the edges Γ_{ij}. The last step consists of determining V_H in each Ω_k so that

$$A_k(V_H,\chi)=0 \quad , \quad \forall\ \chi\epsilon S_h^0(\Omega_k) .$$

Hence the solution of 10) is determined as $V=V_P+V_H$.

The process just described is that required for applying the "action" of the matrix B^{-1} on an arbitrary vector. The local problems may be chosen to "fit" the original operator and so, for example, large discontinuities in the coefficients may be handled without difficulties. The local problems are all independent of each other so that any available parallel computer architecture may be advantageously used. This may prove to be one of the most interesting aspects of the construction of our preconditioner.

To illustrate the performance of the above method we have developed computer programs for model problems in two dimensions. We will present examples which illustrate some of the features of the approach and give some idea of the rates of convergence for the iterative procedures.

The table illustrates the behavior of this type of method as the number of unknowns in a subregion changes. We consider solving problem 1) where Ω is the unit square and L is $-\Delta$. We use piecewise linear functions in triangles and a regular grid with $(1/h-1)^2$ unknowns and partition Ω into sixteen subdomains. The table gives the condition number of the relevant systems as a function of h and a bound on the number of iterations for PCG required to reduce the error by .001. Note that the actual condition number seems to show a $(\ell n\ d/h)^2$ growth instead of the worst case theoretical bound of $(\ell n\ d/h)^3$. The number of iterations required to reduce the error correspondingly grows linearly with $(\ell n\ d/h)$.

h	No. of iterations for .001 reduction	Condition number	$(\log_2 1/h)^2/3.5$	No. of unknowns
1/8	6	3	2.6	49
1/16	8	4.5	4.5	225
1/32	10	7	7	961
1/64	12	10.3	10.3	3969
1/128	14	14	14	16129
1/256	16	18.6	18.3	65025

The method is extremely insensitive to jumps in the coefficients across the boundaries of the subdomains. As a test we considered the above domain decomposition and the variable coefficient problem

$$-\mu(x,y)\ \Delta\ u = f \quad \text{in} \quad \Omega$$
$$u = 0 \quad \text{on} \quad \partial\Omega$$

where μ was piecewise constant on the subregions with constant values as indicated in the following figure.

$\mu=1$	$\mu=100$	$\mu=1$	$\mu=1$
$\mu=1$	$\mu=10$	$\mu=10$	$\mu=10$
$\mu=1$	$\mu=10^6$	$\mu=1$	$\mu=10$
$\mu=1$	$\mu=1000$	$\mu=10^6$	$\mu=10$

The results for the condition number and the number of iterations required for the above problem were within a couple of percent of the results for the constant coefficient problem in the table. Thus these iterative methods will be extremely effective on interface problems even when the coefficients change drastically across the interfaces as long as the interface boundaries align with the subdomain boundaries.

SOME SUPERCONVERGENCE RESULTS FOR MIXED FINITE ELEMENT METHODS FOR LINEAR PARABOLIC PROBLEMS

Mie Nakata and Mary Fanett Wheeler*

Abstract

We consider continuous-time mixed finite element methods with Raviart-Thomas approximating subspaces for linear parabolic problems. Superconvergence results, L_∞ in time and discrete L_2 in space, are derived for both the solution and gradients (velocity).

1. *Introduction*

We consider mixed finite element methods for approximating the pair $(u;p)$ satisfying the parabolic system

$$u(x,y,t) = -a(x,y)\nabla p(x,y,t), \qquad (x,y,t)\in\Omega\times(0,T], \tag{1.1}$$

$$p_t(x,y,t)+\nabla\cdot u(x,y,t)=f(x,y,t), \qquad (x,y,t)\in\Omega\times(0,T], \tag{1.2}$$

with the Neumann boundary condition

$$u\cdot\nu = 0, \qquad (x,y)\in\partial\Omega, \tag{1.3}$$

and initial condition

$$p(x,y,0) = p_0(x,y), \qquad (x,y)\in\Omega. \tag{1.4}$$

Here $\Omega=(0,1)^2$ with the boundary $\partial\Omega$, and ν is the outward normal vector on $\partial\Omega$. We assume that the coefficients $a^i\in W^1_\infty(\Omega)$, where $a=(a^1,a^2)$, and satisfy $0<a_0\le a^i\le a_1$ for some positive real numbers a_0 and a_1. By $a\nabla p$ we mean $(a^1\frac{\partial p}{\partial x}, a^2\frac{\partial p}{\partial y})$. We also assume f is a function such that $f(\cdot,\cdot,\tau)\in L_2(\Omega)$ for each $t=\tau$ and $p_0\in W_3^{r+2}$, r a nonnegative integer to be defined later.

Using Raviart-Thomas approximating spaces [3] in the mixed finite element formulation, we demonstrate superconvergence results for both p and $-a\nabla p$ in the discrete L_2-norm in time. This is an extension of work on elliptic equations by Nakata, Weiser and Wheeler [2] to linear parabolic equations. Superconvergence results for p have been previously derived by Douglas and Roberts [1] for elliptic problems.

The analysis is presented for Neumann boundary conditions; however similar results can be obtained for the Dirichlet problem.

In this paper, for simplicity we have dealt with continuous-time mixed finite element methods but the results can be extended to the discrete case similar to that reported by Weiser and Wheeler [4].

2. *Notation and Formulation of the Mixed Finite Element Method*

*Department of Mathematical Sciences, Rice University, Houston, TX 77251.

We write (z, w) to denote $\int_\Omega z \cdot w \, dx$, the standard L_2-inner product, where $\cdot$ is the dot product. Let $H(div; \Omega) = \{v \in L_2(\Omega)^2 \mid div\ v \in L_2(\Omega)\}$,

$$V = \{v \in H(div; \Omega) \mid v \cdot \nu = 0 \ on\ \partial\Omega\},$$

$$D = L_2(\Omega),$$

and

$$W_p^s(\Omega) = \left\{ f : \Omega \to \mathbf{R} \mid \frac{\partial^m \partial^n}{\partial x^m \partial y^n} f \in L_p(\Omega) \ for\ all\ integers \right.$$

$$\left. \begin{matrix} m \geq 0 \text{ and } n \geq 0 \ such\ that \\ m + n \leq s \ . \end{matrix} \right\} .$$

The norm $||\cdot||_{W_p^s(\Omega)}$ on $W_p^s(\Omega)$ is defined by

$$|| f ||_{W_p^s(\Omega)}^p = \sum_{\substack{n,m \\ n+m \leq s}} || \frac{\partial^m \partial^n}{\partial x^m \partial y^n} f ||_{L_p(\Omega)}^p \ .$$

Let $\delta_x : 0 = x_0 < x_1 < \cdots < x_N = 1$ and $\delta_y : 0 = y_0 < y_1 < \cdots < y_{N_y} = 1$ be partitions of $[0,1]$. Set

$$h = \max_{\substack{i=1,\ldots,N_x \\ j=1,\ldots,N_y}} \{x_i - x_{i-1}, y_j - y_{j-1}\} ,$$

$$V_h^r = V_h^{r,1} \times V_h^{r,2} ,$$

with

$$V_h^{r,1} = M_{0,0}^{r+1}(\delta_x) \otimes M_{-1}^r(\delta_y) ,$$

and

$$V_h^{r,2} = M_{-1}^r(\delta_x) \otimes M_{0,0}^{r+1}(\delta_y) ,$$

and

$$D_h^r = M_{-1}^r(\delta_x) \otimes M_{-1}^r(\delta_y) .$$

Here

$$M_k^s(\delta_x) = \{v \in C^k(I) \mid v_{|I_i} \in P_s(I_i), i = 1, 2, \ldots, N_x\} ,$$

$I = [0,1]$, I_i the ith subinterval and $P_s(I_i)$ the set of polynomials of degree less than or equal to s defined on I_i. We denote by $C^k(I)$ the set of real valued functions defined on I that have k continuous derivatives if $k \geq 0$, and the set of functions if $k = -1$, and $M_{k,0}^s(\delta_x) = M_k^s(\delta_x) \cap H_0^1(I)$. The space V_h^r and D_h^r are usually referred to as Raviart-Thomas spaces, and we note that $div\ V_h^r \subset D_h^r$.

We now motivate the definition of the mixed finite element method. Multiplying (1.1) by $v \in V_h^r$ and integrating by parts we see that at each $t = \tau$,

$$(\frac{1}{a} u, v) - (p, div\ v) = 0, \quad v \in V_h^r . \tag{2.1}$$

Here by $\frac{1}{a} u$ we mean $(\frac{1}{a^1} u^1, \frac{1}{a^2} u^2)$. Multiplying (1.2) by $w \in D_h^r$ and integrating we obtain for $t = \tau$

$$(p_t, w)+(div\ u, w) = (f, w), \qquad w \in D_h^r . \tag{2.2}$$

Trivially we have

$$(p, w) = (p_0, w), \qquad w \in D_h^r,\ t=0 . \tag{2.3}$$

In the mixed finite element method formulation we seek an approximating pair $(U;P)$ where both U and P are differentiable with respect to t; and for each $t=\tau>0$, $(U(\cdot,\cdot,\tau);P(\cdot,\cdot,\tau))\in V_h^r \times D_h^r$ and satisfies

$$(\frac{1}{a} U, v)-(P, div\ u) = 0, \qquad v \in V_h^r, \tag{2.4}$$

$$(P_t, w)+(div\ U, w) = (f, w), \qquad w \in D_h^r, \tag{2.5}$$

with $(U(\cdot,\cdot,0);P(\cdot,\cdot,0))$ satisfying an initial condition to be defined later. The unique existence of such a pair comes from the observation that the arising matrix equation is the initial value problem for a system of ordinary differential equations.

We need to introduce some additional notation. As in [2] we define for a function $f:\Omega\rightarrow \mathbf{R}$

$$|||f|||_x^2 = \sum_{i=1}^{N_x}\sum_{j=1}^{N_y}\sum_{l=1}^{r+2}\sum_{k=1}^{r+1} f(\overline{x}_{il}, y_{jk})^2 h_{i-1} h'_{j-1},$$

$$|||f|||_y^2 = \sum_{i=1}^{N_x}\sum_{j=1}^{N_y}\sum_{l=1}^{r+1}\sum_{k=1}^{r+2} f(x_{il}, \overline{y}_{jk})^2 h_{i-1} h'_{j-1},$$

and

$$|||f|||_x^2 = \sum_{i=1}^{N_x}\sum_{j=1}^{N_y}\sum_{l=1}^{r+1}\sum_{k=1}^{r+1} f(x_{il}, y_{jk})^2 h_{i-1} h'_{j-1},$$

where $\{x_{il}\}(\{\overline{x}_{il}\})$, $\{y_{jk}\}(\{\overline{y}_{jk}\})$ are the $r+1(r+2)$ Gauss points on $[x_{i-1},x_i]([y_{j-1},y_j])$, respectively, and $h_{i-1}=x_i-x_{i-1}$, $h'_{j-1}=y_j-y_{j-1}$. Also for $f:\Omega\times[0,T]\rightarrow\mathbf{R}$

$$||f||^p_{L_p([0,T],||\cdot||)} = \int_0^T ||f||^p(t)\,dt \quad \text{for} \quad 1\le p<\infty ,$$

and

$$||f||_{L_\infty([0,T],||\cdot||)} = \sup_{t\in[0,T]} ||f||(t) ,$$

where $||\cdot||$ is an arbitrary seminorm, e.g., $|||\cdot|||$, $|||\cdot|||_x$, $|||\cdot|||_y$, etc.

Next, we consider the elliptic equation at each $t=\tau$:

$$u = -a\nabla p, \qquad (x,y)\in\Omega , \tag{2.7}$$

$$\nabla\cdot u = f - p_t, \qquad (x,y)\in\Omega . \tag{2.8}$$

Then as before we see that for each $t=\tau$

$$(\frac{1}{a}u, v)-(p, div\ v)=0, \qquad v\in V_h^r, \tag{2.9}$$

$$(div\ u, w) = (f-p_t, w), \qquad w\in D_h^r . \tag{2.10}$$

Let $(W;Z)$ be the approximating solution pair in the mixed finite element method for the elliptic equation, namely

$$(\frac{1}{a}W,v)-(Z,div\ v)=0, \qquad v\in V_h^r, \tag{2.11}$$

$$(div\ W,w)=(f-p_t,w), \qquad w\in D_h^r. \tag{2.12}$$

Note that the solution p for the problem (1.1)-(1.4) is a solution for the problem (2.7)-(2.8) for each $t=\tau$. The L_2-projection of p onto D_h^r is the element $\hat{P}\in D_h^r$ defined by the relation

$$(\hat{P}-p,w)=0, \qquad w\in D_h^r.$$

3. *Superconvergence Results*

In this section we derive superconvergence results. We show that if $(U;P)\in V_h^r\times D_h^r$ satisfies (2.4)-(2.6), then under certain assumptions, for every $\tau\in(0,T]$

$$|||P-p|||(\tau)+|||U^1-u^1|||_x(\tau)+|||U^2-u^2|||_y(\tau)=0(h^{r+2}).$$

This result yields an additional order of h over the usual L_2-norm, $0(h^{r+1})$. The latter result is optimal; i.e., the exponent on h, namely $r+1$, is the largest possible.

We prove the following theorem.

Theorem 3.1 Let $(u;p)$ satisfy (2.1)-(2.3). We assume that $p,p_t\in W_3^{r+2}(\Omega)$, $u^i, u_t^i\in W_\infty^{r+2}(\Omega)$ for $i=1,2$ and $f_t,p_{tt}\in L_2(\Omega)$ for each $t=\tau\in(0,T]$. Let $(U;P)$ and $(W;Z)$ be defined by (2.4)-(2.5) and (2.11)-(2.12) respectively. We also assume $f,p\in W_2^1(0,T)$ for each $(x,y)\in\Omega$, and $(p-Z,1)=(Z_t-\hat{P}_t,1)=0$ for each $t=\tau$. Then there exists a constant C depending on T and a_1 but not on h such that

$$\begin{aligned}&||p-P||_{L_\infty([0,T],|||\cdot|||)}+||u^1-U^1||_{L_\infty([0,T],|||\cdot|||_x)}\\&+||u^2-U^2||_{L_\infty([0,T],|||\cdot|||_y)}\le Ch^{r+2},\end{aligned} \tag{3.1}$$

where $U^i(\cdot,\cdot,0)=W^i(\cdot,\cdot,0)$ and $P(\cdot,\cdot,0)=Z(\cdot,\cdot,0)$.

In order to prove Theorem 3.1 we use a technique established by Wheeler [5]. We consider (2.7)-(2.8) and the approximating solution pair $(W;Z)$ defined by (2.11)-(2.12) and derive estimates of $|||Z-P|||$, $|||W^1-U^1|||_x$ and $|||W^2-U^2|||_y$.

Lemma 3.1 Let $(U;P)$ be defined by (2.4)-(2.5) and $(W;Z)$ by (2.11)-(2.12). Then there exists a constant $C=C(T)$ independent of h such that

$$\begin{aligned}&||Z-P||_{L_2(\Omega)}^2(T)\\&\le C\Big[||Z-P||_{L_2(\Omega)}^2(0)+||(Z-\hat{P})_t||^2_{L_2([0,T],||\cdot||_{L_2(\Omega)})}\Big].\end{aligned}$$

Proof. We first note that Z is differentiable with respect to t since both f and p_t are. Then by adding the term (Z_t,w) to both sides of (2.12) we obtain from (2.11)-(2.12) for each $t=\tau$

$$(\frac{1}{a}W,v)-(Z,div\ v)=0, \qquad v\in V_h^r, \tag{3.2}$$

$$(Z_t,w)+(div\ W,w)=(f,w)+(Z_t-p_t,w), \qquad w\in D_h^r. \tag{3.3}$$

Subtracting (2.4) and (2.5) from (3.2) and (3.3), respectively, we have

$$(\frac{1}{a}(W-U),v)-(Z-P,div\ v)=0, \qquad v\in V_h^r, \tag{3.4}$$

$$(Z_t-P_t,w)+(div(W-U),w)=(Z_t-p_t,w)$$
$$=(Z_t-\hat{P}_t,w), \qquad w\in D_h^r. \tag{3.5}$$

The last equality holds since $\hat{P}_t$ is the L_2-projection of p_t at $t=\tau$. Setting $w=Z-P$ in (3.5) we obtain

$$(Z_t-P_t,Z-P)+(div(W-U),Z-P)=(Z_t-\hat{P}_t,Z-P).$$

Setting $v=W-U$ in (3.4) and substituting the result into the above we deduce that

$$((Z-P)_t,Z-P)+(\frac{1}{a}(W-U),W-U)=((Z-\hat{P})_t,Z-P).$$

Thus,

$$\frac{1}{2}\frac{d}{dt}||Z-P||^2_{L_2(\Omega)}+(\frac{1}{a}(W-U),W-U)$$
$$\le ||(Z-\hat{P})_t||_{L_2(\Omega)}\,||Z-P||_{L_2(\Omega)}$$
$$\le \frac{1}{2}\Big[||(Z-\hat{P})_t||^2_{L_2(\Omega)}+||Z-P||^2_{L_2(\Omega)}\Big],$$

or

$$\frac{d}{dt}||Z-P||^2_{L_2(\Omega)}\le ||(Z-\hat{P})_t||^2_{L_2(\Omega)}+||Z-P||^2_{L_2(\Omega)}.$$

Integrating both sides from $t=0$ to $t=T$ we obtain

$$||Z-P||^2_{L_2(\Omega)}(T)-||Z-P||^2_{L_2(\Omega)}(0)$$
$$\le \int_0^T ||(Z-\hat{P})_t||^2_{L_2(\Omega)}(t)dt+\int_0^T ||(Z-P)||^2_{L_2(\Omega)}(t)dt$$

By Gronwall's inequality, we see that

$$||Z-P||^2_{L_2(\Omega)}(T)$$
$$\le C\Big[||Z-P||^2_{L_2(\Omega)}(0)+||(Z-\hat{P})_t||^2_{L_2([0,T],||\cdot||_{L_2(\Omega)})}\Big]$$

for some constant C which depends on T.

We need another lemma.

Lemma 3.2 Under the same assumption as in Lemma 3.1,

$$||W-U||^2_{L_2(\Omega)}(T)$$
$$\le C_1||W-U||^2_{L_2(\Omega)}(0)+C_2||(Z-\hat{P})_t||^2_{L_2([0,T],||\cdot||_{L_2(\Omega)})},$$

where C_1 and C_2 are constants depending on a_1.

Proof. By the same assumption given for Z, W is also differentiable with respect to t. Thus by differentiating both sides of (3.4) with respect to t we obtain

$$(\frac{1}{a}(W-U)_t, v)-((Z-P)_{t,}\ div\ v) = 0, \qquad v \in V_h^r. \tag{3.6}$$

Setting $v = W-U$ at $t=\tau$ in (3.6) we have

$$\begin{aligned}(\frac{1}{a}(W-U)_t, W-U) &= ((Z-P)_{t,}\ div(W-U)) \\ &= -(div(W-U), div(W-U)) \\ &\quad + ((Z-\hat{P})_t, div(W-U)).\end{aligned}$$

The second equality is obtained by using (3.5) with $w = div(W-U)$. Thus,

$$\frac{1}{2}\frac{d}{dt}\,||\frac{1}{\sqrt{a}}(W-U)||^2_{L_2(\Omega)} + ||div(W-U)||^2_{L_2(\Omega)}$$

$$\leq \frac{1}{2}\left[||(Z-\hat{P})_t||^2_{L_2(\Omega)} + ||div(W-U)||^2_{L_2(\Omega)}\right].$$

Integrating both sides from $t=0$ to $t=T$ we obtain

$$||\frac{1}{\sqrt{a}}(W-U)||^2_{L_2(\Omega)}(T) \leq ||\frac{1}{\sqrt{a}}(W-U)||^2_{L_2(\Omega)}(0)$$

$$+\int_0^T ||(Z-\hat{P})_t||^2_{L_2(\Omega)}(t)dt\,.$$

Thus,

$$||(W-U)||^2_{L_2(\Omega)}(T) \leq C_1||W-U||^2_{L_2(\Omega)}(0)$$

$$+ C_2||(Z-\hat{P})_t||^2_{L_2([0,T],||\cdot||_{L_2(\Omega)})}.$$

Now we are ready to prove Theorem 3.1.

Proof of Theorem 3.1. By the triangle inequality.

$$|||P-p|||(T) \leq |||P-Z|||(T) + |||Z-p|||(T). \tag{3.7}$$

From the elliptic results in [2] we have

$$|||Z-p|||(T) = 0(h^{r+2}). \tag{3.8}$$

By equivalence of norms on D_h^r ($|||\cdot|||$ is a norm on D_h^r),

$$|||P-Z|||(T) \leq const.\ ||P-Z||_{L_2(\Omega)}(T).$$

By setting $P(0) = Z(0)$, we have by Lemma 3.1

$$||P-Z||_{L_2(\Omega)}(T) \leq const.\ ||(Z-\hat{P})_t||^2_{L_2([0,T],||\cdot||_{L_2(\Omega)})}. \tag{3.9}$$

Differentiating (2.9)-(2.10) and (2.11)-(2.12) with respect to t we observe that $(W_t; Z_t)$ is a mixed finite element approximation to $(u_t; p_t)$. Hence we apply the results of [2] to obtain for each $t=\tau$

$$||(Z-\hat{P})_t||_{L_2(\Omega)}(\tau) \leq C\,h^{r+2}, \tag{3.10}$$

where C is a constant independent of h. Combining the results (3.8)-(3.10) we note that

$$|||P-Z|||(T)=0(h^{r+2}). \tag{3.11}$$

We next estimate $|||U^1-u^1|||_x(T)$ and $|||U^2-u^2|||_y(T)$. By the triangle inequality,

$$|||U^1-u^1|||_x(T) \le |||U^1-W^1|||_x(T)+|||W^1-u^1|||_x(T). \tag{3.12}$$

We note from [2] that

$$|||W^1-u^1|||_x(T)=0(h^{r+2}). \tag{3.13}$$

The equivalence of norms on $V_h^{r,1}$ yields

$$|||U^1-W^1|||_x(T) \le const.\ ||U^1-W^1||_{L_2(\Omega)}(T). \tag{3.14}$$

Setting $U(0)=W(0)$ we obtain from Lemma 3.2 that

$$||U^1-W^1||_{L_2(\Omega)}(T)=const.\ h^{r+2}$$

The estimate of $|||U^2-u^2|||_y(T)$ can be obtained similarly. Proof of the theorem now follows.

We remark that for Dirichlet boundary condition, $p=g$ on $\partial\Omega$ we define the finite dimensional space V_h^r as

$$V_h^r=M_0^{r+1}(\delta_x)\otimes M_{-1}^r(\delta_y)\times M_{-1}^r(\delta_x)\otimes M_0^{r+1}(\delta_y).$$

Thus we have at each $t=\tau$

$$(\frac{1}{a}u,v)-(p,div\ v)=-\int_{\partial\Omega} g(v\cdot\nu), \qquad v\in V_h^r,$$

$$(p_t,w)+(div\ u,w)=(f,w), \qquad w\in D_h^r.$$

Noting that the result of [2] holds also for elliptic problems with Dirichlet boundary conditions an analogous argument applies as before. Extensions to $\Omega\subset\mathbf{R}^n$, $n>2$ are also straightforward.

REFERENCES

1. Douglas, J. Jr.; Roberts, Jean:
   Global estimates for mixed methods for second order elliptic equations.
   To appear in Math. of Comp. 44 (1985).

2. Nakata, M.; Weiser, A.; Wheeler, M.F.:
   Some superconvergence results for mixed finite element methods for elliptic problems on rectangular domains.
   To appear in MAFELAP Proceedings (1985).

3. Raviart, P.A.; Thomas, J.M.:
   A mixed finite element method for 2nd order elliptic problems in Mathematical Aspects of the Finite Element Method. *Lecture Notes in Mathematics*, Springer-Verlag, Rome (1975), Heidelberg (1977).

4. Weiser, A.; Wheeler, M.F.:

On convergence of block-centered finite differences for elliptic problems.
To appear.

5. Wheeler, M.F.:
A priori L_2 estimates for Galerkin approximations to parabolic partial differential equations.
SIAM J. Numer. Anal. 10, 723-759 (1973).

NODAL METHODS FOR THE NUMERICAL SOLUTION OF PARTIAL DIFFERENTIAL EQUATIONS

J.P. Hennart
IIMAS-UNAM
Apartado Postal 20-726
01000 México, D.F.
(MEXICO)

1.- INTRODUCTION

Nodal methods appeared in nuclear engineering in the second half of the 1970's decade, in numerical reactor calculation and especially in neutronics calculations, where all kinds of partial differential equations (PDEs) are to be solved. Static group diffusion calculations have to do with coupled systems of elliptic PDEs, parabolic PDEs are found in connection with space-time kinetics problems, while hyperbolic ones are characteristic of neutron transport problems. Basically, nodal methods were developed in the USA around J.J. Dorning, then at the University of Illinois in Urbana-Champaign, and A.F. Henry at the MIT in Boston, as well as in the Federal Republic of Germany by people like M.R. Wagner, W. Werner and their collaborators. Some classical references are Finnemann et al. [1], Langenbuch et al. [2,3], Shober et al. [4]. The review papers by Frohlich [5], Dorning [6] and Wagner and Koebke [7] should also be mentioned.

In a sense, nodal methods are intermediate between the finite difference method (FDM) and the finite element method (FEM). Of the FEM, they retain its accurate aspect, thanks to the (implicit or explicit) use of some piecewise continuous approximation of the solution, mostly polynomials but sometimes also more complicated functions (in the so-called "analytical" nodal methods), in a piecewise way. As the FDM, they lead to sparse and well-structured systems of algebraic equations, at least over not too irregular grids, typically of the union of rectangles type. Strictly speaking, they are thus fast solvers, with many vectorial and parallel aspects.

In numerical reactor calculation, the nodal calculation itself is only one part of the global procedure, as it is normally preceded by a preprocessing operation and followed by a postprocessing one. As it turns out, man-made light water reactor cores are tremendously heterogeneous and cell calculations are first performed to get homogenized properties (the "coefficients" in the corresponding equations) over a fairly coarse mesh. The next stage is the nodal or "coarse mesh" calculation itself, which tries to capture the main behavior of the unknown function (here the "neutron flux") by providing for instance very accurate estimates of its mean value over each cell, block, element or "node". A last stage is finally needed, whereby the fine details of the neutron flux are revealed cell by cell by some dehomogenization technique.

As a matter of fact, there are many situations taken from other fields of application, where everything boils down to the coarse-mesh or nodal calculation, which was the central stage in numerical reactor calculation. These situations are characterized by the combination of the following facts: first of all, the standard approach has always been the FDM over regular meshes of the union of rectangles type, maybe simply by inertia of the industry (finite differences have been present for a long time and there is no particular incentive to invest a lot of money in rewriting most of the existing software; rectangular cells or blocks have always been used and nobody knows or wants to know what a triangle is (and certainly not what a tetrahedron is in three dimensions!)); moreover the properties of the medium being modeled are not well-known, so that a mean value or constant per cell is probably the best guess for the corresponding coefficient. This combination is typical of flow in porous media (underground hydrology, oil reservoir simulation, etc.). Air pollution modeling is another example of application and our list is certainly not exhaustive.

2.- BASIC FORMULATION OF NODAL SCHEMES

In this section, we want to illustrate some of the simplest nodal schemes and relate them to more conventional finite element ones. At this point, it is interesting to quote Wagner and Koebke's recent paper [7]: speaking of modern nodal methods, they characterize them by three distinct features, the first of which being that the unknowns chosen are node-averaged or surface-averaged ones (in this case, the neutron flux over the cell and the corresponding neutron (net or partial) "current" through the edges). To clarify these points, it is convenient now to consider some model problem, namely a typical second order elliptic PDE:

$$Lu \equiv \vec{\nabla} \cdot p\vec{\nabla}u + qu = f \quad \text{on} \quad \Omega \qquad , \qquad (1a)$$

where the unknown u is subject to boundary conditions on $\Gamma=\bar{\Omega}-\Omega$ that we will take for the sake of simplicity of the Dirichlet type:

$$u = 0 \quad \text{on} \quad \Gamma \quad . \tag{1b}$$

In the neutronics situations, u would be the neutron flux, while $\vec{v}=-p\vec{\nabla}u$ would be the neutron (net) current.

At this stage, it is convenient to introduce some notation. In two dimensions, P_k will denote the space of polynomials of total degree k

$$P_k \equiv \{x^a y^b \mid 0 \leq a + b \leq k\} \quad , \tag{2}$$

and $Q_{k,\ell}$, the space of polynomials of degree k in x and ℓ in y

$$Q_{k,\ell} \equiv \{x^a y^b \mid 0 \leq a \leq k, \ 0 \leq b \leq \ell\} \quad . \tag{3}$$

In particular, $Q_k \equiv Q_{k,k}$. Let us moreover introduce the following linear functionals associated with the reference cell $C \equiv [-1,+1] \times [-1,+1]$ (to which any other cell can be mapped by an affine diagonal transformation)

$$m_C^{ij}(u) \equiv \int_C P_{ij}(x,y) \; u(x,y)dxdy/N_i \cdot N_j \quad , \tag{4a}$$

and with its edges L,R,B, and T (for Left, Right, Bottom, and Top, respectively)

$$m_L^i(u) \equiv \int_L P_i(y) \; u(-1,y)dy/N_i \quad ,$$

$$m_R^i(u) \equiv \int_R P_i(y) \; u(+1,y)dy/N_i \quad ,$$

$$m_B^i(u) \equiv \int_B P_i(x) \; u(x,-1)dx/N_i \quad ,$$

and

$$m_T^i(u) \equiv \int_T P_i(x) \; u(x,+1)dx/N_i \quad , \tag{4b}$$

where $N_i = 2/(2i+1)$ is a convenient normalization factor while the P_i's are the normalized Legendre polynomials over $[-1,+1]$ with $P_{ij} \equiv P_{ij}(x,y) = P_i(x)P_j(y)$.

The simplest nodal scheme, known in the nuclear engineering literature as QUABOX [2], can be described as follows: over a given node, u is approximated by u_h, defined in terms of its mean value over the node ($m_C^{oo} \equiv m_C^{oo}(u_h)$) and of its mean values on the edges of the node (m_L^o, m_R^o, m_B^o, and m_T^o). In the original paper [2], these mean values are replaced by values at the centers of gravity of the cell and of its edges. Using instead mean values, and later cell and edge moments, turns out to be more practical with respect to the "Patch Test" considerations which will be made later. The derivation of the QUABOX scheme is then based on the following "physical" considerations: first, a balance equation over the current element Ω_e is expressed as

$$\int_{\Omega_e} (Lu_h - f)d\vec{r} = 0 \quad . \tag{5a}$$

If the given cell and its neighboring cells on the left, right, bottom and top share the same mean flux on their common edge, mean flux continuity through the edges is automatically ensured. The condition (suggested by the physics of the problem) which will thus be imposed will be a mean current continuity condition through the corresponding edges, i.e.

$$\int_{L,R,B,\text{ or }T} (\vec{v}_{h+} - \vec{v}_{h-}) \cdot \vec{1}_n \, ds = 0 \quad , \tag{5b}$$

where $\vec{v}_h \equiv -p\vec{\nabla}u_h$, while $\vec{1}_n$ is some unit normal to the edge considered. These conditions (5a & 5b) provide the basic equations relating the node-averaged neutrons flux and the corresponding edge-averaged neutron currents through the left, right, bottom and top faces of the node considered.

As in [2], the approximation u_h to the neutron flux u over a given node can be expressed in terms of its mean value over the node and of its mean values over the corresponding edges. More generally, the unknown function, here the neutron flux u, is approximated by a function u_h, described by a set D of degrees of freedom, consisting of linear functionals acting on some space of functions S, cell by cell.

Corresponding to D = $\{m_L^o, m_R^o, m_B^o, m_T^o, \text{ and } m_C^{oo}\}$, let S be given by $Q_{2,0} \cup Q_{0,2}$. Clearly card D = dim S, a necessary but not sufficient condition for the "D-unisolvence of S", which means that there is one and only one member of S satisfying the "interpolation" conditions contained in D. To prove the D-unisolvence of S, there are two basic techniques: the first one consists in exhibiting the members of S satisfying the interpolation conditions of D (if they exist, they are unique since dim S = card D); otherwise, we must show that if all the degrees of freedom in D are set equal to zero, then the only member of S which will satisfy them is identical to zero.

For the QUABOX scheme, we have

THEOREM 1:

$S = Q_{2,0} \cup Q_{0,2}$ is D-unisolvent, where $D = \{m_L^o, m_R^o, m_B^o, m_T^o$, and $m_C^{oo}\}$.

PROOF:

With card D = dim S, the basis functions corresponding to m_L^o, m_R^o, m_B^o, m_T^o, m_C^{oo} are

$$u_L^o = -\frac{1}{2}(P_{10} - P_{20})$$
$$u_R^o = +\frac{1}{2}(P_{10} + P_{20})$$
$$u_B^o = -\frac{1}{2}(P_{01} - P_{02})$$
$$u_T^o = +\frac{1}{2}(P_{01} + P_{02})$$
$$u_C^{oo} = P_{00} - P_{20} - P_{02} \quad .$$

It is clearly tempting to try using the corresponding local approximation to the neutron flux in a more conventional finite element scheme, where the original equation (1) is first written in a so-called weak form. Namely, the problem of finding a solution to (1) is replaced by the following one:

Find $u \in H_0^1(\Omega)$ such that

$$\int_\Omega (p\vec{\nabla}u \cdot \vec{\nabla}v + quv - fv)\, d\vec{r} = 0 \quad , \quad \forall \quad v \in H_0^1(\Omega) \quad , \qquad (6)$$

where $H_0^1(\Omega)$ is the standard Sobolev space. This problem being of the same difficulty as the original one, some discretization must be attempted by which instead of considering all of $H_0^1(\Omega)$, some finite dimensional subspace of it is selected, let say S_h, and the final formulation becomes:

Find $u_h \in S_h \subset H_0^1(\Omega)$ such that

$$\int_\Omega (p\vec{\nabla}u_h \cdot \vec{\nabla}v_h + qu_h v_h - fv_h)\,d\vec{r} = 0 \quad , \quad \forall\ v_h \in S_h \subset H_0^1(\Omega) \quad . \tag{7}$$

The problem with the basis functions of the QUABOX scheme is that they do not satisfy the "conformity" condition $S_h \subset H_0^1(\Omega)$, as only the mean value of u_h is continuous across the interfaces but not u_h itself. The standard way to face this problem, or "variational crime" according to Strang [8], consists in replacing $a(u,v)$ in (7) by $a_h(u,v)$ defined as

$$a_h(u,v) = \sum_e \int_{\Omega_e} (p\vec{\nabla}u \cdot \vec{\nabla}v + quv)\,d\vec{r} \quad . \tag{8}$$

Not all the nonconforming finite elements will, when used in conjunction with (8), give rise to approximations u_h of u converging to it, as the mesh size is refined. As a matter of fact, we shall hereafter refer to a consistency test known as the Patch Test [8,9] which remains a useful practical tool for testing nonconforming elements, even if it sometimes fails in pathological situations [10]. At its lowest order, the Patch Test is satisfied when the mean value of the approximation is continuous across interfaces, which is exactly what our choice of parameters is doing. If we exactly perform the quadratures in (8), we get a set of algebraic equations to solve, which is still quite different from the ones we obtained previously by pure physical considerations. The corresponding primal (u is the basic unknown) nonconforming FEM is what we use to call a Mathematical Nodal Method (MNM), valid on its own, to stress the fact that it is distinct from the Physical Nodal Method (PNM) obtained hereabove on a more physical-basis. To get the PNM from the MNM, a second variational crime is needed, i.e. the quadratures in (8) are to be performed numerically. It turns out that a numerical quadrature of the tensor product Radau type leads to that result: namely, the QUABOX scheme, in its physical formulation is identical to a primal nonconforming FEM, when this numerical quadrature is used to obtain the final algebraic equations. This was first shown in [11]. We note that the quadrature scheme is not standard, as different tensor product rules combining the left or right Radau points with the bottom and top Radau points are used, depending on which product of basis functions is integrated. More details can be found in [12, 13]. A complete numerical analysis of QUABOX leading to error bounds in H^1-

norm of $0(h)$, and subsequently to error bounds in L^2-norm of $0(h^2)$ following classical Aubin-Nitsche arguments, can be found in [14], h being the maximum diameter of the nodes. In [11], a whole family of nodal schemes of the so-called sum or Σ form [2,3] was in fact considered. With k a nonnegative integer ($k \in \mathbb{N}$), let

$$D \equiv D_k = \{m_L^o, m_R^o, m_B^o, m_T^o, \quad m_C^{io}, i = 0,\dots,k, \quad m_C^{oj}, j = 1,\dots,k\}, \quad \text{card } D_k = 2k+5,$$

and

$$S \equiv S_k = Q_{k+2,0} \cup Q_{0,k+2}, \quad \dim S_k = 2k+5,$$

$$\forall\, k \in \mathbb{N}.$$

We have

THEOREM 2:

S_k is D_k-unisolvent $\forall\, k \in \mathbb{N}$.

PROOF:

With card D_k = dim S_k = 2k+5 the basis functions corresponding to the different edge and cell moments are

$$u_L^o = \tfrac{1}{2}(-1)^{k+1}\,(P_{k+1,0} - P_{k+2,0}),$$
$$u_R^o = \tfrac{1}{2}\,(P_{k+1,0} + P_{k+2,0}),$$
$$u_B^o = \tfrac{1}{2}(-1)^{k+1}\,(P_{0,k+1} - P_{0,k+2}),$$
$$u_T^o = \tfrac{1}{2}\,(P_{0,k+1} + P_{0,k+2}),$$

$$u_C^{oo} = P_{00} - P_{k+m(0),0} - P_{0,k+m(0)} ,$$
$$u_C^{io} = P_{i0} - P_{k+m(i),0} , \quad i = 1,\dots,k,$$
$$u_C^{oj} = P_{0j} - P_{0,k+m(j)} , \quad j = 1,\dots,k,$$

where m(i) (resp. m(j)) = 1 or 2 and is such that i and k+m(i) (resp. j and k+m(j)) have the same parity.

We note that the case k = 0 corresponds to the QUABOX scheme, and we have

THEOREM 3:

The nodal scheme of the sum or Σ form associated to $k \in \mathbb{N}$, in either its mathematical form or its physical form, will provide convergence of O(h) in H^1 norm of $O(h^2)$ in L^2 norm.

SKETCH OF THE PROOF:

The proof is largely based on the techniques of Chapter 4 of Ciarlet's book [15]. With the MNM, the only variational crime is due to the nonconformity of the basis functions ($S_h \not\subset H_0^1(\Omega)$). Following the second Strang lemma [15, p. 210], the error in some H^1-like norm consists of two terms: the first one depends on the approximation properties of S_h cell by cell, while the second one depends on the mean continuity conditions between adjacent cells. Since the Patch Test is only passed at its lowest order (the mean value or moment of zeroth order only is continuous on the edges of the cells) and that moreover only $P_1 \subset S_k$ but never P_ℓ, $\ell > 1$, for any k, convergence orders of O(h) in H^1 norm and of $O(h^2)$ in L^2 norm only can be achieved as confirmed numerically in [11,14]. If a PNM is considered instead of a MNM, the effect of numerical integration must also be taken into account. Since the quadrature scheme is of the tensor product Radau type with k+2 quadrature points in each direction, it is exact for the members of Q_{2k+2}: consequently [15], the errors introduced by numerical quadrature, even when k=0, will not modify the convergence orders mentioned hereabove.

In the following, we shall call "finite element of nodal type" or "nodal element" any finite element in which the basic degrees of freedom are edge and (or) cell moments, instead of the classical values and (or) derivatives at discrete points. If the (nonnegative integer) index k is associated to a nodal element of a given family, we shall say that this family of nodal elements climbs correctly in order if L^2 error bounds of $O(h^{k+2})$ can be expected. In that sense, the family of nodal schemes we considered in [11] do not climb correctly in order. In [12], we develop a family of nodal schemes climbing correctly in order and where D_k and S_k are given by

$$D_k = \{u_L^i, u_R^i, u_B^i, u_T^i,\ i = 0,\dots,k\ ;\ u^{ij},\ i,j = 0,\dots,k\}$$

$$,\ \text{card } D_k = (k+1)(k+5)$$

and

$$S_k = Q_{k+2,k} \cup Q_{k,k+2}\ ,\quad \dim S_k = (k+1)(k+5)\ ,$$

$$\forall\ k \in \mathbb{N}.$$

Again, we note that the QUABOX scheme corresponds to $k = 0$, and we have

THEOREM 4:

S_k is D_k - unisolvent $\forall\ k \in \mathbb{N}$.

PROOF: (see [12])

With card D_k = dim S_k = (k+1) (k+5), the basis functions corresponding to the different edge and cell moments are:

$$u_L^i = \frac{1}{2}(-1)^{k+1} (P_{k+1,i} - P_{k+2,i}) \quad , \quad i = 0,\ldots,k,$$

$$u_R^i = \frac{1}{2} (P_{k+1,i} + P_{k+2,i}) \quad , \quad i = 0,\ldots,k,$$

$$u_B^i = \frac{1}{2}(-1)^{k+1} (P_{i,k+1} - P_{i,k+2}) \quad , \quad i = 0,\ldots,k,$$

$$u_T^i = \frac{1}{2} (P_{i,k+1} + P_{i,k+2}) \quad , \quad i = 0,\ldots,k,$$

and

$$u_C^{ij} = P_{ij} - P_{k+m(i),j} - P_{i,k+m(j)} \quad , \quad i,j = 0,\ldots,k,$$

where m(i) (resp. m(j)) = 1 or 2 and is such that i and k+m(i) (resp. j and k+m(j)) have the same parity.

Moreover

THEOREM 5:

The above nodal schemes, either in their mathematical or physical form, exhibit convergences of $0(h^{k+1})$ in H^1 norm and of $0(h^{k+2})$ in L^2 norm.

SKETCH OF THE PROOF:

Similar to the one given for Theorem 3, with the differences that here $P_{k+1} \subset S_k$, $\forall\ k \in \mathbb{N}$ and that moreover moments of order up to k are common between two neighboring cells so that a Patch Test of order k is passed. Finally if a PNM is considered instead of an MNM, by expressing cell balance with respect to $x^a y^b$, $0 \leq a, b \leq k$, as well as current moments continuity on the edges up to order k (the flux moments continuity on the edges up to order k being ensured by the initial selection of degrees of freedom), it can be shown [12,13] that this PNM can be derived from the corresponding MNM if a nonstandard tensor product Radau quadrature rule is used, as in the case of the previous Sum - or Σ - schemes. As for Theorem 3, the error introduced by this numerical quadrature does not modify the convergence orders proved for the MNM, as numerical results do confirm [16,17].

Let us now mention some practical aspects about the numerical implementation of the above nodal schemes.

As mentioned in [12], the couplings between horizontal (L and R) and vertical (B and T) components are very weak and suggest ADI-like schemes for the numerical solution of the corresponding algebraic equations. Going back to [7], it is easy to realize that such ADI-schemes are directly proposed via a procedure called Transverse Integration (TI): TI consists in integrating the original PDE between x_i and x_{i+1} or between y_j and y_{j+1}, to get transverse-integrated one-dimensional equations. The spatial dependence of the transverse leakage term which appears in these one-dimensional equations is approximated by a parabolic fit: in a sense, this is a way to "satisfy" a higher order Patch Test, as we directly do it in our approach to nodal schemes. In [7], TI seems to be an indispensable ingredient of nodal schemes: it is not in fact and the original contributions of Langenbuch et al. [2,3] did not use it. This is true at least for static and dynamic diffusion (i.e. elliptic and parabolic) problems. In neutron transport (i.e. hyperbolic) problems, TI is always used and the original equations are transverse-integrated, after eventual multiplication by some weight, in other words some transverse moments of the original equations are always taken, which is not the case in the static and dynamic diffusion situation (see e.g. [18]).

Recently, the nodal schemes we mentioned were connected to extensions or enhancements of mixed-hybrid finite elements following Arnold and Brezzi [19]. In [13] in particular, we showed (with some minor mistakes) the the nodal elements mentioned hereabove are extensions "à la Arnold-Brezzi" of mixed-hybrid finite elements using the standard Raviart-Thomas-Nedelec (RTN) elements [20,21]. The original equation is first written as a system of first order equations relating the flux and the current. In the mixed finite element approximation of such a system of equations, approximations to both flux and currents are looked for in different functional spaces, namely L^2 and H(div). If the approximate current is to be in H(div), the moments of its normal component to the interfaces between different elements must be continuous. Such a conformity condition (in H(div)) can be relaxed by the use of Lagrange parameters, which turn out to give us information about the corresponding moments of the dual variable, namely the flux. If such edge information is combined with the cell information already provided by the original scheme, what we get is as shown in [13], equivalent to the direct use of the finite elements described in [12], with Radau quadrature and TI. The mixed-hybrid approach is however much more efficient numerically speaking: most of the basic operations can be performed in parallel, as the flux and the current can be calculated cell by cell, the only coupled system of algebraic equations being the one relating the Lagrange parameters, which can as a matter of fact be solved ADI-wise. Finally the combination of cell information with edge information given by the Lagrange parameters to provide an enhanced or extended flux is also a cell by cell operation. Numerical experiments wich such schemes will be described in [22].

In the original papers on nodal schemes [2,3], the emphasis was on final algebraic systems relating mean values (or values at the center of gravity) for a given node and its four neighbors (in 2D). Such an approach is typically finite differences oriented. Actually, a finite element approach would normally eliminate the cell parameters by some static condensation technique and finally come up with a system of algebraic equations relating only the edge parameters. It is in fact possible, by nonstandard numerical quadratures, to get systems of five-points (or block five-points) finite difference-like schemes as shown in [13]. Such finite difference schemes are of the block-centered type and they do perform better than the finite differences of the mesh-centered type as they harmonically (and not arithmetically) average the coefficient p in Eq. (1a). This is known to be much better when p is highly heterogeneous [23].

The general problems of building in a constructive way any finite, element of nodal type was recently solved in [24]: the original question was, given a set of degrees of freedom D_k, namely some cell and (or) edge moments, how to find a corresponding space S_k of basic monomials $x^a y^b$ (in 2D for instance) unisolvent for such D_k, and having nice enough approximation properties as "climbing correctly in order". In [24], we propose a constructive algorithm to do so, which is able in all the previously known situations to reproduce the corresponding S_k's. We can moreover generate new families of nodal elements, in particular nodal elements being extensions "à la Arnold-Brezzi" of the mixed finite elements proposed recently by Brezzi, Douglas, and Marini [25]. Some numerical experiments with these new nodal elements are reported in [22].

3.- SOME OTHER APPLICATIONS

In the previous section, we basically described nodal schemes as they are used in connection with PDEs of elliptic type, and in fact (with slight modifications) of parabolic type. In this last section, we want to speak of applications to linear hyperbolic equations, namely the discrete ordinate transport equations in x-y geometry.

$$\mu u_x + \nu u_y + \sigma u = S \qquad , \qquad (9)$$

where u is the neutron angular flux in the direction (μ,ν), σ is some physical coefficient (a "cross-section"), while S is a "source" term, including scattering, fission and possibly external sources. Assuming as before that the spatial domain is the union of rectangular cells of size $\Delta x \times \Delta y$ belonging to the intersection of I vertical slices and J horizontal ones, we can write Eq. (9) over such a node in dimensionless variables (x and y again) as

$$\frac{2\mu}{\Delta x} u_x + \frac{2\nu}{\Delta y} u_y + \sigma u = S \qquad , \qquad (10)$$

where $(x,y) \in [-1,+1]\times[-1,+1]$.

Here in contradiction with what happened for static and dynamic diffusion problems (i.e. elliptic and parabolic PDEs), the standard approach is always through "transverse integration". Namely, (k+1), $k \in \mathbb{N}$, succesive transverse moments of (10) are taken with respect to the y variable, to obtain [18]

$$\frac{2\mu}{\Delta x} \frac{dm_y^\ell(u;x)}{dx} + \sigma m_y^\ell(u;x) = m_y^\ell(S;x) - \frac{2\nu}{\Delta y} L_y^\ell(u;x) \qquad , \qquad \ell = 0,\dots,k \qquad , \qquad (11)$$

where

$$m_y^\ell(v;x) \equiv \int_{-1}^{+1} P_\ell(y) v(x,y) dy / N_\ell \qquad . \qquad (12)$$

In (11), L_y^ℓ is a transverse leakage depending on u (and x) at the top and bottom of the cell and on m_y^ℓ, m=0,...,ℓ-1, and we have for instance

$$L_y^0(u;x) \equiv (u(x,+1)-u(x,-1))/2 \qquad (13.a)$$

$$L_y^1(u;x) \equiv 3(u(x,+1)+u(x,-1))/2 - m_y^0(u;x), \qquad (13.b)$$

etc.. Similar equations can be obtained by taking transverse moments in the x direction and clearly compatibility conditions must be satisfied which actually require that m_x^i and m_y^j commute. This is automatically ensured by our approach where u is uniquely defined in terms of some edge and (or) cell moments.

Usually equations (11) (and their counterparts in the y direction) are solved by a "diagonal sweeping" method: if (μ,ν) is a direction in the first quadrant, the first cell to be looked at is the cell of the bottom left, let say the (1,1) cell, of the mesh. Knowing by the boundary conditions, the neutron angular flux (and the corresponding moments if applicable) incoming at the left and bottom of this cell, equations (11) are solved to get the corresponding quantities at the right and top boundaries. The procedure is then repeated to produce the same answer for the cells on the next diagonal ((2.1) and (1.2)), and so on. In [26], we experiment with a novel scheme, by which Eqs. (11) are solved in parallel, for the I horizontal and J vertical slices constituting the domain of interest. This approach has interesting aspects: if the classical "diagonal sweeping" does have some vectorization features which make it interesting for the current modern supercomputers, it is in fact basically sequential as the successive diagonals of the mesh are to be inspected one by one. In our approach, parallelism is evident at each step. More details are offered in [26].

REFERENCES

[1] Finneman, H., Bennewitz, F., and Wagner, M.R., "Interface current techniques for multidimensional reactor calculations", Atomkernenergie 30, 123-128 (1977).

[2] Langenbuch, S., Maurer, W., and Werner, W., "Coarse-mesh flux-expansion method for the analysis of space-time effects in large light water reactor cores", Nucl. Sci. Engng. 63, 437-456 (1977).

[3] Langenbuch, S., Maurer, W., and Werner, W., "High-order schemes for neutron kinetics calculations based on a local polynomial approximation", Nucl. Sci. Engng. 64, 508-516 (1977).

[4] Shober, R.A., Sims, R.N., and Henry, A.F., "Two nodal methods for solving time-dependent group diffusion equations", Nucl. Sci. Engng. 64, 582-592 (1977).

[5] Frohlich, R., "Summary discussion and state of the art review for coarse-mesh computational methods", Atomkernenergie 30, 152-158 (1977).

[6] Dorning, J.J., "Modern coarse-mesh methods - A development of the 70's" in Computational Methods in Nuclear Engineering, Vol. 1, pp. 3.1-3.31, American Nuclear Society, Williamsburg, Virginia (1979).

[7] Wagner, M.R. and Koebke, K., "Progress in nodal reactor analysis", Atomkernenergie 43, 117-126 (1983).

[8] Strang, G. and Fix, G.J., An Analysis of the Finite Element Method, Prentice-Hall, Englewood Cliffs, New Jersey (1973).

[9] Gladwell, I. and Wait, R., Eds., A Survey of Numerical Methods for Partial Differential Equations, Clarendon Press, Oxford (1979).

[10] Stummel, F., "The limitations of the patch test", Int. J. Numer. Methods Eng. 15, 177-188 (1980).

[11] Fedon-Magnaud, C., Hennart, J.P., and Lautard, J.J., "On the relationship between some nodal schemes and the finite element method in static diffusion calculations" in Advances in Reactor Computations, Vol. 2, pp. 987-1000, American Nuclear Society, Salt Lake City, Utah (1983).

[12] Hennart, J.P., "A general family of nodal schemes", SIAM J. on Scientific and Statistical Computing 7, 264-287 (1986).

[13] Hennart, J.P., "Nodal schemes, mixed-hybrid finite elements and block-centered finite differences", INRIA Rapports de Recherche, No. 386, 59 p. (1985).

[14] Fedon-Magnaud, C., Etude Theorique de Quelques Methodes Nodales de Resolution de L'Equation de Diffusion - Tests Numeriques, Note CEA-N-2358, 94 p. (1983).

[15] Ciarlet, P., The Finite Element Method for Elliptic Problems, North-Holland, Amsterdam (1978).

[16] Hennart, J.P., "A general finite element framework for nodal methods", in The Mathematics of Finite Elements and Applications,

pp. 309-316, Whiteman, J.R., Ed., Academic Press, London (1985).

[17] Del Valle, E., Hennart, J.P., and Meade, D., "Finite element formulations of nodal schemes for neutron diffusion and transport problems", Nucl. Sci. Engng. 92, 204-211 (1986).

[18] Hennart, J.P., "A general approach to nodal schemes in numerical transport theory", Comunicaciones Técnicas, Serie Naranja: Investigaciones, No. 382, 24 p., IIMAS-UNAM (1985).

[19] Arnold, D.N., and Brezzi, F., "Mixed and nonconforming finite element methods: implementation, postprocessing, and error estimates", M^2AN 19, 7-32 (1985).

[20] Raviart, P.A. and Thomas, J.M., "A mixed finite element method for 2nd order elliptic problems", in Lecture Notes in Mathematics, 606, pp. 292-315, Springer-Verlag, Berlin (1977).

[21] Nedelec, J.C., "Mixed finite elements in R3", Numer. Math. 35, 315-341 (1980).

[22] Del Valle, E., Hennart, J.P., and Meade, D., In Preparation.

[23] Bensoussan, A., Lions, J.L., and Papanicolau, G., Asymptotic Analysis of Periodic Structures, North-Holland, Amsterdam (1978).

[24] Hennart, J.P., Jaffre, J., and Roberts, J.E., "A constructive method for deriving finite elements of nodal type", To Appear.

[25] Brezzi, F., Douglas, J., Jr., and Marini, L.D., "Two families of mixed finite elements for second order elliptic problems", Numer. Math. 47, 217-235 (1985).

[26] Del Valle, E., Filio, C.L., and Hennart, J.P., In Preparation.

SINGULAR PERTURBATION PROBLEMS IN SEMICONDUCTOR DEVICES

Franco BREZZI (*°), Antonio CAPELO (*), Luisa Donatella MARINI (°)

1. INTRODUCTION.

The aim of this paper is to focus on some particular features of boundary value problems arising in the study of reverse-biased semiconductor devices. Roughly speaking, we analyze the model as a singular perturbation problem, where the perturbation parameter is related to the temperature. Our conjecture is that, when the temperature goes to zero, the problem becomes a free-boundary problem, which corresponds to the well known (and widely used) assumption of total depletion. This can be proved on simple one-dimensional models, but the proof in more general cases is not straightforward. In the real-life problems, the temperature is usually the "room temperature" and it is unreasonable to let it go to zero. However the solution of the problem at room temperature exibits internal layers which are already very sharp, so that the total depletion assumption is often used with success to compute approximate solutions. We believe that a better understanding of the process that goes from the full model to the one with total depletion should be helpful for more efficient numerical computations.

An outline of the paper is the following. In section 2 we present a tipical set of equations, in two dimensions, and we point out the singular perturbation parameter. In section 3 we analyze a (very) simplified one-dimensional case and we show that, as the singular perturbation parameter goes to zero, the problem becomes a free-boundary problem (which is, in this case, very easy to solve). In section 4 some computational results on the one-dimensional case are presented.

Acknowledgments: We wish to express our thanks to P.L. Lions, to A. Savini and especially to N. Nassif for the helpful discussions.

(*) Dipartimento di Meccanica Strutturale dell'Università di Pavia.

(°) Istituto di Analisi Numerica del Consiglio Nazionale delle Ricerche (Pavia).

Work partially supported by MPI 60%.

2. THE TWO-DIMENSIONAL PROBLEM.

From the mathematical point of view, the problem can be described as follows. We are given a (smooth) domain Ω in $\mathbb{R}^2$ and we look for functions $\psi(x)$, $n(x)$, $p(x)$ $(x=(x_1,x_2))$ defined on Ω such that

$$
(2.1) \quad \begin{cases} \operatorname{div}(\varepsilon \operatorname{grad} \psi) = -q(-n+p+N_d-N_a) \\ \operatorname{div}(q\mu_n n \operatorname{grad} \psi - q D_n \operatorname{grad} n)=0 \\ \operatorname{div}(q\mu_p p \operatorname{grad} \psi + q D_p \operatorname{grad} p)=0 \\ + \text{ boundary conditions} . \end{cases}
$$

In (2.1), ε, q, μ_n, D_n, μ_p, D_p are given positive constants (this is already a simplification: ε might depend on x and μ_n, D_n, μ_p, D_p might be nonlinear functionals depending on ψ, n, p). The functions $N_d(x)$ and $N_a(x)$ are given nonnegative functions. We may assume, for the sake of simplicity, that both $N_d(x)$ and $N_a(x)$ have the form

$$
N_d(x)=\bar{N}_d\,\chi_d(x), \quad N_a(x)=\bar{N}_a\,\chi_a(x),
$$

where $\chi_d(x)$ and $\chi_a(x)$ are the characteristic functions of two given (smooth) subsets of Ω, say Ω_d and Ω_a, with $\Omega_d \cap \Omega_a = \emptyset$ and $\bar{\Omega}_d \cup \bar{\Omega}_a = \bar{\Omega}$.

Without giving many details on the physical background, we briefly indicate the physical meaning of the data and, roughly, their order of magnitude:

ε = permittivity ($\simeq 10^{-10}$F/m)

q = charge of the electron ($\simeq 10^{-19}$C)

N_d = donor's doping ($\bar{N}_d \simeq 10^{22}\,m^{-3}$)

N_a = acceptor's doping ($\bar{N}_a \simeq 10^{22}\,m^{-3}$)

μ_n, μ_p = mobility of electrons and holes, respectively

D_n, D_p = diffusion coefficients for electrons and holes, respectively

$\left[D_n/\mu_n = D_p/\mu_p \;\; (=kT/q) \simeq 10^{-2}V\right]$.

As far as the unknowns are concerned:

$\psi(x)$ = electric potential

$n(x)$, $p(x)$ = concentration of free electrons and holes, respectively.

As we can see, the orders of magnitude involved have very big variations. It is therefore convenient to do some scaling. In order to simplify the computation, let us (brutally) assume that $\bar{N}_d = \bar{N}_a$; then (2.1) can be reduced to:

$$
(2.2) \quad \begin{cases} \Delta \tilde{\psi} = \tilde{n} - \tilde{p} - \chi_d + \chi_a \\ \operatorname{div}(\tilde{n} \operatorname{grad} \tilde{\psi} - \lambda \operatorname{grad} \tilde{n}) = 0 \\ \operatorname{div}(\tilde{p} \operatorname{grad} \tilde{\psi} + \lambda \operatorname{grad} \tilde{p}) = 0 \\ + \text{ boundary conditions} . \end{cases}
$$

Here $\tilde{\psi}$, $\tilde{n}$, $\tilde{p}$ are the scaled version of ψ, n, p, and the only parameter left is, in this case:

$$\lambda = \frac{\varepsilon D_n}{\mu_n q \bar{N}_d} = \frac{\varepsilon D_p}{\mu_p q \bar{N}_a} \simeq 10^{-15} m^2.$$

It is clear that (2.2) should be regarded as a singular perturbation problem; in this respect, one should, in particular, analyze the limit of $\tilde{\psi}$, $\tilde{n}$, $\tilde{p}$ as λ goes to zero. We believe that, in particular, the so called "total depletion assumption" (often used by engineers in studying the reverse-biased devices) could be justified, from the mathematical point of view, as the limit of (2.2) for $\lambda \to 0$. Unfortunately, this, at present, is only a guess. Nevertheless, in the next section, we present a proof of this conjecture in a simplified one-dimensional case.

3. A ONE-DIMENSIONAL CASE.

For the sake of simplicity, we consider here the one-dimensional version of (2.2) and we assume moreover that $\Omega_d = \emptyset$. We note that the latter assumption is not really a restriction if one studies the reverse-biased case. Hence our model problem is now:

$$(3.1) \quad \begin{cases} u'' = 1 - v & \text{in }]0,a[\\ (v u' + \lambda v')' = 0 & \text{in }]0,a[\\ u(0) = 0, \; u(a) = b, \; v(0) = 1, \; v(a) = 0 \end{cases}$$

and we assume that $b>0$ (this corresponds to the choice of the reverse-biased case). We now prove that (3.1) has at least one solution for every $\lambda > 0$.

Theorem 3.1: *For every* $\lambda>0$ *problem* (3.1) *has at least one solution* (u,v) *which moreover satisfies* $0 \leqslant v \leqslant 1$.

Proof: We set

$$K = \{\phi \mid \phi \in L^\infty(0,a),\; 0 \leqslant \phi \leqslant 1\};$$

for every $\phi \in K$ we define $u = u_\phi$ as the solution of

$$(3.2) \quad \begin{cases} u'' = 1 - \phi & \text{in }]0,a[\\ u(0) = 0, \; u(a) = b. \end{cases}$$

Then we consider $w = w_\phi$ as the solution of

$$(3.3) \quad \begin{cases} (\lambda w' + w u_\phi')' = 0 \\ w(0) = 1, \; w(a) = 0. \end{cases}$$

We have, explicitely,

$$(3.4) \quad w(x) = \left[e^{-u(x)/\lambda} \int_x^a e^{u(t)/\lambda} dt \right] \Big/ \int_0^a e^{u(t)/\lambda} dt.$$

Finally we set:

$$(3.5) \quad F_\phi(x) := P_{[0,1]} w_\phi := \begin{cases} w_\phi(x) & \text{if } 0 \leqslant w_\phi(x) \leqslant 1 \\ 0 & \text{if } w_\phi(x) < 0 \\ 1 & \text{if } w_\phi(x) > 1 . \end{cases}$$

It is an easy matter to check that $F_\phi(x)\in W^{1,\infty}(0,a)\cap K$. Therefore the mapping $\phi\to F_\phi$ is compact from K to K. Hence it has a fixed point by the Leray-Schauder theorem. Let $v\in K$ be such that $F_v = v$. The proof is then concluded if we show that $P_{[0,1]}w_v = w_v$ (that is, if $0\leq w_v\leq 1$). Clearly $w_v\geq 0$ from (3.4); assume now that there exists an interval $[r,s]$ such that $w = w_v\geq 1$ in $[r,s]$ with $w(r) = w(s) = 1$. In $[r,s]$ we have $P_{[0,1]}w\equiv 1$. Since $v = F_v = P_{[0,1]}w$ we have $v\equiv 1$ in $[r,s]$ and then (from (3.2)) $u_v'' = 0$, that is $u_v' = k =$ constant in $[r,s]$. Hence from (3.3) we get

$$(3.6)\qquad \begin{cases} \lambda w'' + k w' = 0 & \text{in }]r,s[\\ w(r) = w(s) = 1 \end{cases}$$

which implies $w\equiv 1$ in $[r,s]$. ■

We want now to study the behaviour of the solution (u,v) (provided by theorem 3.1) as λ goes to zero. From now on we shall denote such a solution by (u_λ, v_λ) to underline the dependence on λ. We shall always assume that

$$(3.7)\qquad 0\leq v_\lambda \leq 1$$

as it is allowed by theorem 3.1.

The following result is an easy consequence of equations (3.1):

Proposition 3.2: *If* (u_λ, v_λ) *is a solution of* (3.1) *which belongs to* $(L^2(0,a))^2$, *then both* u_λ *and* v_λ *are analytic.*

The next theorem provides some further information on the behaviour of (u_λ, v_λ).

Theorem 3.3: *If* (u_λ, v_λ) *is a solution of* (3.1) *satisfying* (3.7) *then* v_λ *is strictly decreasing.*

Proof: It is clear that v_λ cannot take the value 0 at an internal point x_0. Indeed, if this were the case v_λ would satisfy in the interval $[x_0, 1]$ the linear equation

$$(3.8)\qquad \lambda v_\lambda'' + v_\lambda' u_\lambda' + v_\lambda u_\lambda'' = 0$$

with the boundary conditions $v_\lambda = 0$ at x_0 and 1. Hence $v_\lambda\equiv 0$ in $[x_0, 1]$, which is impossible since v_λ is analytic. The proof is concluded by noting that v_λ cannot have a positive minimum due to the maximum principle. ■

We need now some a priori bounds on u_λ, v_λ independent of λ. From theorem 3.3 and the first equation of (3.1) we immediately obtain the following:

Proposition 3.4: *There exists a constant* $C>0$ *such that for every* $\lambda>0$ *and for every solution* (u_λ, v_λ) *of* (3.1) *satisfying* (3.7):

$$(3.9)\qquad \|u_\lambda\|_{W^{3,1}} + \|v_\lambda\|_{W^{1,1}} \leq C.$$

From proposition 3.4 and Sobolev embedding theorems we obtain that there exists a subsequence (still denoted (u_λ, v_λ)) such that

$$(3.10)\quad \begin{cases} u_\lambda \to u_0 & \text{in } H^2(0,a) \\ v_\lambda \to v_0 & \text{in } L^2(0,a)\,. \end{cases}$$

Let us write the second equation of (3.1) as

$$(3.11)\qquad \lambda v' + vu' = C_\lambda = \text{constant.}$$

A simple computation shows that the boundary conditions $v_\lambda(0)=1$, $v_\lambda(a)=0$ imply

$$(3.12)\qquad C_\lambda = -\lambda \Big/ \int_0^a e^{u_\lambda(t)/\lambda}\,dt.$$

Since $u_0(a)=b>0$ we have

$$(3.13)\qquad \lim_{\lambda\to 0} C_\lambda = 0\,.$$

Moreover we have

$$(3.14)\qquad \lambda v' \to 0 \quad \text{in } L^1(0,a)$$

and, since $u_\lambda' \to u_0'$ in L^∞ and $v_\lambda \to v_0$ in L^2,

$$(3.15)\qquad u_\lambda' v_\lambda \to u_0' v_0 \quad \text{in } L^2(0,a).$$

Hence we can take the limit of (3.11) for $\lambda\to 0$. We have

$$(3.16)\qquad u_0' v_0 = 0.$$

Taking also the limit of the first equation of (3.1) for $\lambda\to 0$, we have that (u_0,v_0) satisfy the following conditions:

$$(3.17)\quad \begin{cases} u_0'' = 1 - v_0 \\ u_0' v_0 = 0 \\ u_0(0)=0,\ u_0(a)=b \\ 0 \leqslant v_0 \leqslant 1. \end{cases}$$

We summarize these results in the following theorem:

Theorem 3.5: *From any sequence* $\{(u_\lambda,v_\lambda)\}_\lambda$ *of solutions of* (3.1) *it is possible to extract a subsequence (still noted* $\{(u_\lambda,v_\lambda)\}_\lambda$*) such that* u_λ *converges strongly in* $H^2(0,a)$ *and* v_λ *converges strongly in* $L^2(0,a)$. *Moreover, setting* $u_0 := \lim u_\lambda$, $v_0 := \lim v_\lambda$ *we have that* (u_0,v_0) *satisfies* (3.17).

As we shall see in a moment (3.17) do not characterize u_0, v_0 in a unique way, so that we have to add to (3.17) the information that $v_0 = \lim v_\lambda$ with (u_λ,v_λ) solution of (3.1).

Theorem 3.6: *If* (u_0,v_0) *is a solution of* (3.17) *with* $v_0 \in L^2(0,a)$, *then there exist* $\alpha,\beta\in[0,a]$, $\alpha\leqslant\beta$, *such that* $v_0 = \chi_{[\alpha,\beta]}$ *a.e. (here* $\chi_{[\alpha,\beta]}=1$ *if* $\alpha\leqslant x\leqslant\beta$ *and zero otherwise).*

Proof: We start by proving that if $u_0'(x_1)=u_0'(x_2)=0$ with $x_1<x_2$ then $u_0'\equiv 0$ in $[x_1,x_2]$. This is easy: from $u_0''=1-v_0\geqslant 0$ we have that u_0' is nondecreasing. It follows that, if $u_0'(x_1)=u_0'(x_2)=0$, then $v_0=1-u_0''=1$ in $[x_1,x_2]$. Set now $\alpha=\min\{x \mid u_0'(x)=0\}$ and $\beta=\max\{x \mid u_0'(x)=0\}$. We have already shown that $v_0=1$ in $[\alpha,\beta]$. Clearly $u_0'\neq 0$ outside $[\alpha,\beta]$ and hence $v_0=0$ outside $[\alpha,\beta]$ (because $v_0u_0'=0$). ■

On the opposite, it is clear that, for any pair α,β of points in $[0,a]$ such that $\alpha\leq\beta$ and $(a-\beta)^2-\alpha^2=2b$ we have a solution of (3.17) by setting $v_0=\chi_{[\alpha,\beta]}$ and $u_0''=1-v_0$ with $u_0(0)=0$, $u_0(a)=b$. Hence (3.17) has infinitely many solutions, if we ask $v_0\in L^2(0,a)$. However we know that, if $v_0 = \lim v_\lambda$, then v_0 must be a function of bounded variation, and v_0' must be a negative measure. An easy checking shows now that this leaves us with the two only possibilities $v_0^1\equiv 0$ (for $x>0$) and $v_0^2=\chi_{[0,\xi]}$ with $(a-\xi)^2=2b$ (whenever such a ξ exists: for $2b\geq a^2$, the choice v_0^1 is already the only one possible). The next lemma excludes one of the two:

Lemma 3.7: *If* $\{(u_\lambda,v_\lambda)\}_\lambda$ *is a sequence of solutions of* (3.1), *and* $u_0 = \lim u_\lambda$, $v_0=\lim v_\lambda$ *(in* H^2 *and* L^2 *respectively), then* $u_\lambda'(0)\geq 0$.

Proof: Assume $u_0'(0)<0$. Then $u_\lambda'(0)<u_0'(0)/2$ for λ, say, smaller than λ_0. From (3.11) we have $\lambda v_\lambda'(0)=C_\lambda-u_\lambda'(0)$. From (3.13) we have then $\lambda v_\lambda'(0)> -u_0'(0)/4$ for λ small enough, which would give $v_\lambda'(0)>0$, in contrast with theorem 3.3. ■

Lemma 3.7 implies that the solution $v_0^1\equiv 0$ (for $x>0$) is acceptable (as limit of v_λ) only if the corresponding u_0 (given by $u_0''=1$, $u_0(0)=0$, $u_0(a)=b$) is nonnegative. This is clearly the case iff $2b\geq a^2$.

We have proved the following result:

Theorem 3.8: *Let* $\{(u_\lambda,v_\lambda)\}_\lambda$ *be a sequence of solutions of* (3.1), *and assume that* $u_0=\lim u_\lambda$ *(in* H^2*) and* $v_0=\lim v_\lambda$ *(in* L^2*). Then* (u_0,v_0) *satisfy* (3.17) *and we have:*

$$\text{(3.18)}\quad \begin{cases} i)\ \textit{for}\ 2b\geq a^2 & v_0(x)=0 \quad \textit{for}\quad x>0 \\ ii)\ \textit{for}\ 2b< a^2 & v_0(x)=\chi_{[0,\xi]},\quad (a-\xi)^2=2b. \end{cases}$$

Clearly (3.18) characterizes (u_0,v_0) in a unique way as a solution of (3.17).

In view of the two-dimensional case, it might be worthwile noting that the following "version" of (3.17) has a unique solution (the one given by (3.18)):

$$\text{(3.19)}\quad \begin{cases} \textit{Find}\ v_0\in BV(0,a)\ \textit{and}\ u_0\in H^2(0,a)\ \textit{such that} \\ u_0''(x)=1-v_0(x) \quad \textit{in}\ ]0,a[\\ v_0(x)u_0'(x)=0 \quad \textit{in}\ ]0,a[\\ u_0(0)=0,\ u_0(a)=b \\ \lim\limits_{x\to 0} v_0(x)=1 \quad \textit{if}\ u_0'(0)<0. \end{cases}$$

Finally we note that (3.17) can actually be seen as a *free-boundary problem*. Indeed from theorem 3.6 we have that (3.17) can also be written as

$$\text{(3.20)}\quad \begin{cases} \textit{Find}\ u_0\in H^2(0,a)\ \textit{and}\ \Omega\ \textit{open subset of}\ ]0,a[\ \textit{such that} \\ u_0''(x)=1-\chi_\Omega \quad \textit{in}\ ]0,a[\\ u_0(0)=0,\ u_0(a)=b \\ u_0'=0 \quad \textit{on}\ ]0,a[\cap\partial\Omega \quad (\textit{free-boundary condition}). \end{cases}$$

It is clear that (3.20) has infinitely many solutions. However one may easily check, with the same argument as before, that a unique solution is left if we add the "boundary conditions on χ_Ω":

$$(3.21)\quad \begin{cases} \textit{if } u_0'(0)<0 \textit{ then } 0\in\partial\Omega, \\ \textit{if } u_0'(a)>0 \textit{ then } a\notin\partial\Omega. \end{cases}$$

4. NUMERICAL RESULTS.

The following pictures show the behaviour of $u_\lambda(x)$ and $v_\lambda(x)$ for different values of λ: namely a)$\equiv\lambda=1.$; b)$\equiv\lambda=10^{-3}$; c)$\equiv\lambda=10^{-6}$; d)$\equiv\lambda=10^{-18}$. The results were obtained by discretizing (3.1) using a mesh of 31 nodes for a=1 and b=1/8. For λ less than 10^{-5} the shock in $v_\lambda(x)$ was captured within one mesh interval.

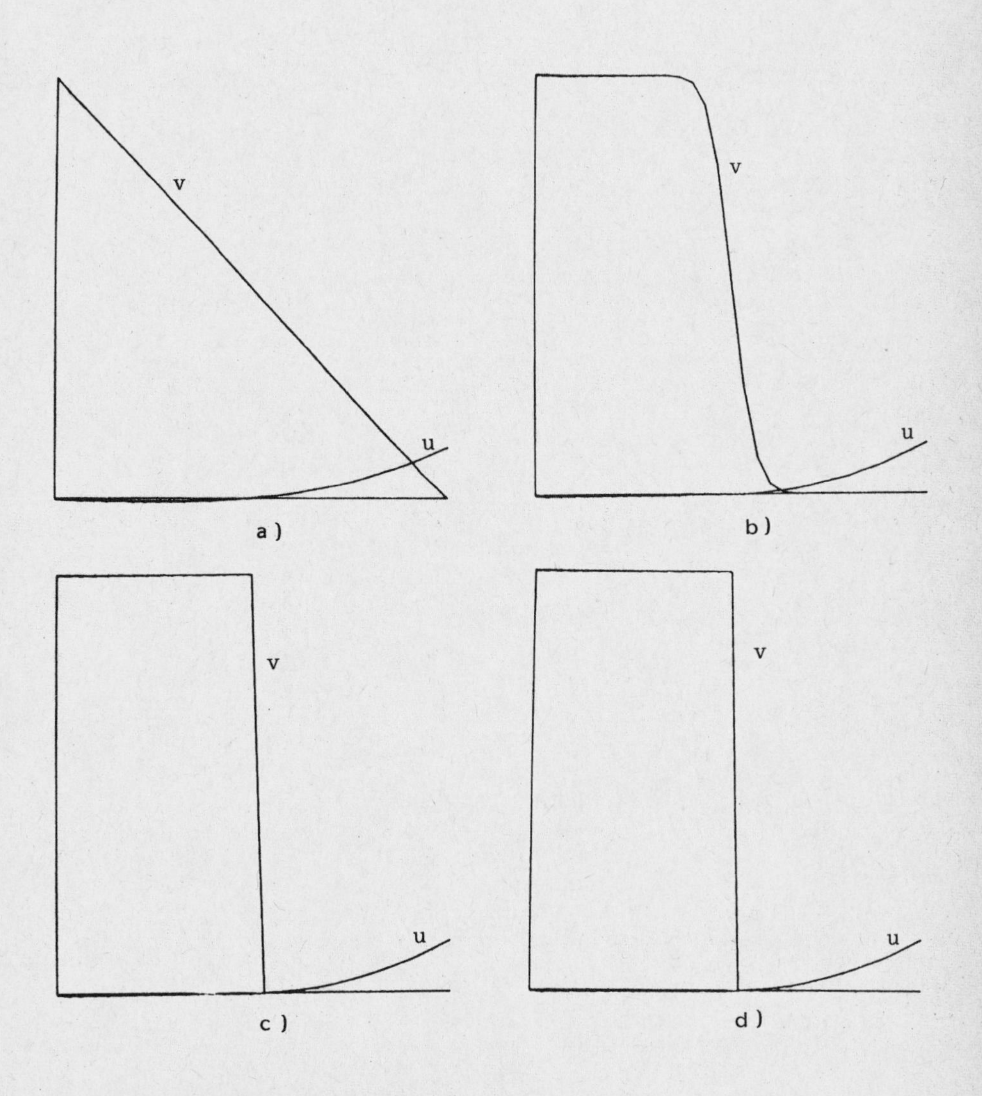

References

R. Bank, D. Rose, W. Fichtner, Numerical methods for semiconductor device simulation, *IEEE Trans. Electron Devices*, 30(1031-1041)1983.

B. Browne, J. Miller, Eds., *Numerical Analysis of Semiconductor Devices*, Boole, Dublin, 1979.

B. Browne, J. Miller, Eds., *Numerical Analysis of Semiconductor Devices and Integrated Circuits*, Boole, Dublin, 1981.

W. Fichtner, D. Rose, R. Bank, Semiconductor device simulation, *IEEE Trans. Electron Devices*, 30(1018-1030)1983.

C. Hunt, N. Nassif, On a variational inequality and its approximation, in the theory of semiconductors, *SIAM J. Numer. Anal.*, 12(938-950)1975.

R. Lyon-Caen, *Diodes, Transistors, and Integrated Circuits for Switching Systems*, Academic Press, New York, 1968.

L. Marini, A. Savini, Accurate computation of electric field in reverse biased semiconductor devices. A mixed finite element approach, *COMPEL*, 3(123-135)1984.

M. Mock, On equations describing steady-state carrier distributions in a semiconductor device, *Comm. Pure Appl. Math.*, 25(781-792)1972.

M. Mock, *Analysis of Mathematical Models of Semiconductor Devices*, Boole, Dublin, 1983.

N. Nassif, K. Malla, Formulation mathématique du comportement de quelques semi-conducteurs au moyen d'une inégalité quasi variationnelle (IQV), *C. R. Ac. Sci. Paris*, 294(79-82)1982.

N. Nassif, K. Malla, Étude de l'existence de la solution d'une inégalité quasi variationnelle apparaissant dans la théorie des semi-conducteurs, *C. R. Ac. Sci. Paris*, 294(119-122)1982.

N. Nassif, K. Malla, Simulation numérique d'un semi-conducteur polarisé en sens inverse au moyen d'une inégalité quasi variationnelle, *C. R. Ac. Sci. Paris*, 294(345-348)1982.

STABILITY OF CAPILLARY WAVES ON DEEP WATER

Benito Chen
IIMAS-UNAM
Apdo. Postal 20-726
01000 México, D.F.

P.G.Saffman
Applied Mathematics 217-50
California Institute of Technology,
Pasadena, Ca. 91125

ABSTRACT

The stability of periodic capillary waves of permanent form on deep water to three dimensional disturbances is studied using numerical methods.

1. INTRODUCTION.

In the last few years there has been a renewed interest in steady, periodic, finite amplitude waves on deep water. Most of the work has been done on gravity waves: Longuet-Higgins and Fox [7] calculated waves of almost greatest height, Chen and Saffman [2] showed that sufficiently steep waves are not unique and there is bifurcation into new classes of two-dimensional waves of permanent form. The stability of two-dimensional gravity waves to two-dimensional perturbations was done by Longuet-Higgins [6], [7]. The stability to three-dimensional perturbations was studied by McLean et al [9] and in more detail by McLean [8].

For capillary waves, Crapper [4] found an exact, analytic two-dimensional solution for waveheights up to the maximum. In this paper we determine the stability boundaries and growth rates for three-dimensional disturbances to steady two-dimensional capillary waves. This is done numerically. Neutral stability curves are also given.

Chen and Saffman [1], [3] studied two-dimensional capillary-gravity waves of finite amplitude and found a very complex bifurcation and limit point structure. It is expected that the study of the stability of these waves will produce some interesting results. We are starting some work in this direction.

2. STABILITY.

Consider inviscid, irrotational waves of permanent form on deep water under the influence of surface tension. Let the two dimensional waves move in the positive x direction with speed c on water otherwise it rest. We superimpose a uniform stream moving in the negative x direction with speed c to make the flow steady. The positive z axis is pointing upward. To study the stability of the steady waves we add a small time dependent disturbance.

The governing equations are

$$\nabla^2 \phi(x,y,z,t) = 0 \qquad -\infty < z < \eta(x,y,t) \qquad (2.1)$$

$$\phi \sim - c\, x \qquad \text{as} \quad z \to -\infty \qquad (2.2)$$

$$\eta_t + \phi_x \eta_x + \phi_y \eta_y = \phi_z \qquad \text{on} \quad z = \eta(x,y,t) \qquad (2.3)$$

$$\phi_t - T\left(\frac{1}{R_1} + \frac{1}{R_2}\right) + \frac{1}{2} \nabla\phi \cdot \nabla\phi = F \quad \text{on } z = \eta(x,y,t) \qquad (2.4)$$

Here the free surface is $z = \eta(x,y,t)$, the potential is $\phi(x,y,z,t)$. T is the surface tension, F is Bernoulli's constant and the principal radii of curvature are given by

$$\frac{1}{R_1} + \frac{1}{R_2} = \frac{\eta_{xx}(1+\eta_y^2)+\eta_{yy}(1+\eta_x^2)-2\,\eta_{xy}\,\eta_x\eta_y}{(1+\eta_x^2+\eta_y^2)^{3/2}}$$

Without loss of generality we take the wavelength $\lambda=2\pi$, the surface tension T=1.

Crapper [4] found exact, steady two dimensional solutions to the above equations, by giving x and z as functions of the potential. For small amplitudes his solutions can be inverted and written as

$$z = \bar{\eta}(x) = \sum_1^\infty A_j \cos j\,x \qquad (2.5a)$$

$$\phi = \bar{\phi}(x,z) = -c\,x + \sum_1^\infty B_j \sin j\,x\, e^{jz} \qquad (2.5b)$$

We study the stability of (2.5) to infinitesimal three dimensional disturbances of the form

$$\eta^1 = e^{-i\sigma t}\, e^{i(px+qy)} \sum_{j=-\infty}^{\infty} a_j\, e^{ijx} \qquad (2.6a)$$

$$\phi^1 = e^{-i\sigma t}\, e^{i(px+qy)} \sum_{j=-\infty}^{\infty} b_j\, e^{ijx}\, e^{[(p+j)^2+q^2]^{1/2} z} \qquad (2.6b)$$

Since we want disturbances that remain finite as $x,y \to \infty$, p and q are real. But apart from that they are arbitrary. The physical disturbance is the real part of (2.6). The equations for (2.6) are

$$\sum_{j=-\infty}^{\infty} \{\bar{\phi}_{xz}\,\bar{\eta}_x - \bar{\phi}_{zz} + i(p+j)\,\bar{\phi}_x\} a_j\, e^{ijx}$$

$$+ \sum_{j=-\infty}^{\infty} (i(p+j)\bar{\eta}_x - [\;]^{1/2}) b_j\, e^{ijx}\, e^{[\;]^{1/2} z} = i\,\sigma \sum_{j=-\infty}^{\infty} a_j\, e^{ijx}$$

$$\text{on } z=\bar{\eta} \qquad (2.7)$$

$$\frac{T}{(1+\bar{\eta}_x^2)^{3/2}} \sum_{j=-\infty}^{\infty} \{(p+j)^2+q^2(1+\bar{\eta}_x^2)+3i(p+j)\ \frac{\bar{\eta}_x\ \bar{\eta}_{xx}}{1+\bar{\eta}_x^2}\}\ a_j\ e^{ijx}$$

$$+(\bar{\phi}_x\ \bar{\phi}_{xz} + \bar{\phi}_z\ \bar{\phi}_{zz}) \sum_{j=-\infty}^{\infty}\ a_j\ e^{ijx}$$

$$+ \sum_{j=-\infty}^{\infty} \{i\ \bar{\phi}_x(p+j)+\ \bar{\phi}_z[(p+j)^2+q^2]^{1/2}\}b_j\ e^{ijx}\ e^{[\ ]^{1/2}z} =$$

$$+\ i\ \sigma \sum_{j=-\infty}^{\infty}\ b_j\ e^{ijx}\ e^{[\ ]^{1/2}z} \quad \text{on } z=\bar{\eta} \tag{2.8}$$

Here [ ] represents $(p+j)^2+q^2$. (2.7)-(2.8) is a generalized eigenvalue problem for the eigenvalue σ and the eigenfunction $\{a_j,b_j\}_{i=-\infty}^{\infty}$

If σ is real then the disturbance is stable; if σ is complex then the disturbance or its complex conjugate is unstable.

Equations (2.7)-(2.8) can be solved immediately when the undisturbed wave has h/λ, the ratio of wavehight to wavelength, equal to zero. The eigenpairs are

$$\sigma_n^{\pm}(p,q)=-(p+n)\ \pm\ ((p+n)^2+q^2)^{3/4} \tag{2.9a}$$

$$\eta_n^1(p,q)\ =\ e^{-i\sigma_n^{\pm}t}\ e^{i[(p+n)x+qy]} \tag{2.9b}$$

which represent infinitesimal capillary waves on top of a flat surface. The ± sign gives the direction of propagation of the perturbation. All the eigenvalues (2.9a) are real.

To have instability we need a complex eigenvalue and its complex conjugate and this can only happen when two real eigenvalues coalesce. For small h/λ instability may happen near the points (p,q) where two of the eigenvalues for zero amplitude are equal

$$\sigma_{n_1}^{\pm}\ (p,q)\ =\ \sigma_{n_2}^{\pm}(p,q) \qquad n_1 \neq n_2 \text{ or signs unequal} \tag{2.10}$$

But (2.10) may be true without implying instability for $h/\lambda \neq 0$, since the two eigenvalues may separate. See figure 1.

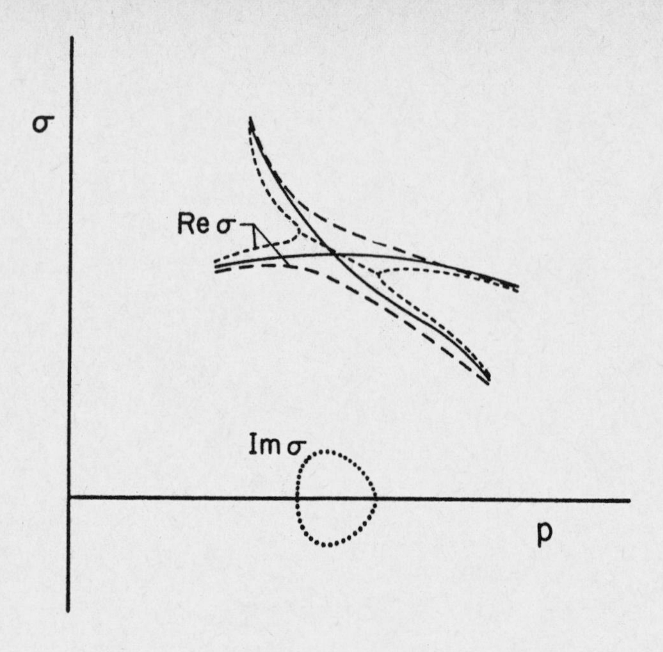

Figure 1.- Sketch of σ vs. p for given q to show possible effect of finite wave height on a linear resonance. The solid line shows dependence for $h/\lambda = 0$. Long dashed line shows stable behavior; short dashed line shows the real part of σ and dotted line the imaginary part for unstable behavior.

In (2.10) we can add any integer to p, so only the difference between n_1 and n_2 is relevant. Following McLean [9] we name the two classes of solutions according to the difference being even or odd, Class I and Class II.

$$\sigma_m^+(p,q) = \sigma_{-m}^+(p,q) \;, \qquad m \geqslant 1 \qquad (2.11a)$$

or

$$[(p+m)^2+q^2]^{3/4} - [(p-m)^2+q^2]^{3/4} = 2m \qquad m \geqslant 1$$

We also get

$$\sigma_0^+(0,0) = \sigma_0^-(0,0) \tag{2.11b}$$

$$\sigma_1^+(0,0) = \sigma_{-1}^-(0,0) \tag{2.11c}$$

Class II

$$\sigma_m^+(p,q) = \sigma_{-m-1}^+(p,q) \qquad m \geqslant 0 \tag{2.12a}$$

or

$$[(p+m)^2+q^2]^{3/4} - [(p-m)^2+q^2]^{3/4} = 2m+1 \qquad m \geqslant 0$$

and also $\sigma_0^+(p,q) = \sigma_{-1}^-(p,q)$ (2.12b)

Curves (2.11) and (2.12) and curves where σ_m is zero are plotted in figures 2.a and 2.b. The class I curves are symmetrical with respect to p=0, and the class II and neutral stability curves with respect to $p=\frac{1}{2}$. All are symmetrical with respect to q=0.(2.11a) and (2.12a) give curves that go to ∞ as p goes to ∞. (2.11b) and (2.11c) give the point at the origin and (2.12b) is a closed curve.

We used numerical methods to find the stability boundaries (2.6a) and (2.6b) were truncated to j from-M to M. The implicit function theorem and Cauchy-Riemann relations were used to obtain the necessary derivatives of Crapper's solution. The 4M+2 unknowns $\{a_j, b_j\}$ j=-M,..,M are chosen to satisfy (2.7)-(2.8) at 2M+1 points equally spaced between two adjacent crests. In this way we obtain a generalized eigenvalue problem of order 4M+2 of the form

$$Au = \sigma Bu \tag{2.13}$$

with $u=[a_{-M},\ldots,a_M,b_{-M},\ldots,b_M]^T$ and A and B complex matrices depending on p,q and the steady solution. The L Z algorithm was used to solve (2.13). For M=16, 34 equations, it took about two minutes to calculate all the eigenvalues σ and about one minute more to get all the eigenfunctions on a VAX 11/750, using double precision. For M=32 it took 18 minutes to calculate all the eigenvalues and six more to also get the eigenfunctions. An "alternative" is to use Newton's method.

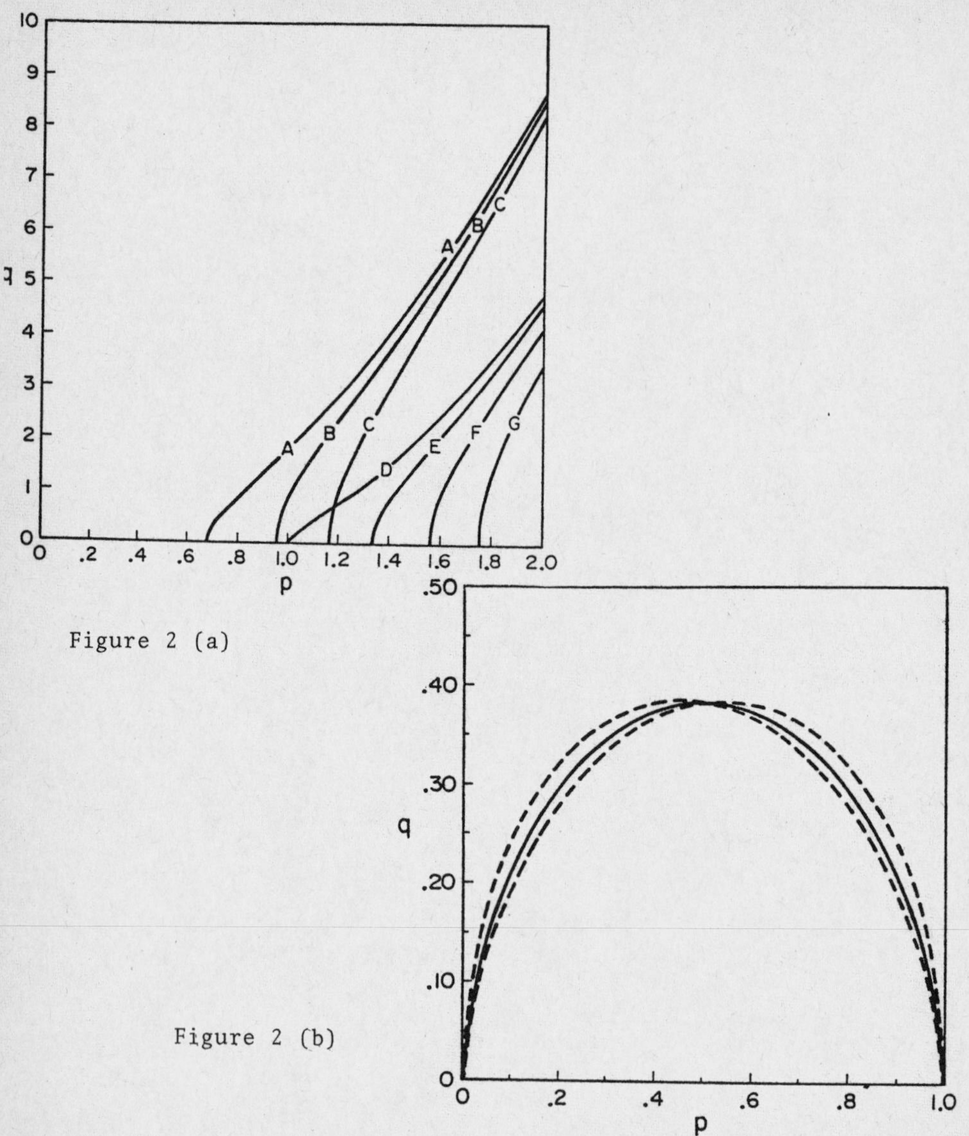

Figure 2 (a)

Figure 2 (b)

Capillary waves.- Curves for Class I and Class II resonances and stationary modes for $h/\lambda = 0$ in the p-q plane. (a) A, Class I m=1; B, Class I m=2; C, Class I, m=3; D, Class II, m=0 (eq 2.12a); E, Class II m=1; F, Class II m=2; G, Class II m=3. (b) Solid line is Class II m=0 (eq 2.12b); dashed lines are stationary disturbances. Wave numbers of resonant modes are p+m and p-m for Class I and p+m and p-m-1 for Class II.

To check the effect of the truncation, some of the calculations were repeated increasing M until the relevant eigenvalues converged and the neglected terms in the corresponding eigenvectors were $\mathcal{O}(10^{-8})$. For h/λ = .03 M=8 gave the relevant eigenvalues to four significant figures. For h/λ = .06 and h/λ = .09, M=16 was used.

The numerical methods were tested by calculating some of the results given by McLean [9] for gravity waves. Also by solving (2.7)-(2.8) for $h/\lambda=0$ and comparing with (2.9). The agreement was very good.

The stability boundaries and neutral stability curves calculated are plotted in figures 3.a and 3.b for h/λ=.03 and .09. The"circular" region near the origin is the expansion of (2.11b) and (2.11c) as h/λ increasis from zero, and corresponds to the Lighthill, Benjamin-Feir, Whitham instability. The narrow banded region of instability corresponds to the class II resonance given by (2.12b). For the values of h/λ given above the maximum growth rate is located at $p=\frac{1}{2}$, q=.383 and is shown by a plus sign. The growth rate is $\mathcal{O}(h/\lambda)$. In contrast the two dimensional growth rate (q=0) is $\mathcal{O}(h/\lambda)^2$, so that for h/λ big enough the two dimensional instability should dominate.

The resonance curves (2.11a) and (2.12a) do not give rise to instabilities as h/λ increases since the eigenvalues separate and stay real.

The method for studying the stability of capillary waves fails for $h/\lambda > .1165$, well below the maximum value, h/λ = .730. The reason is that we are studying the problem in the physical plane and the transformation of Crapper's solution to this plane has a singularity. This singularity moves below the level of the crest for $h/\lambda > .1165$ and the Fourier series (2.5) diverges.

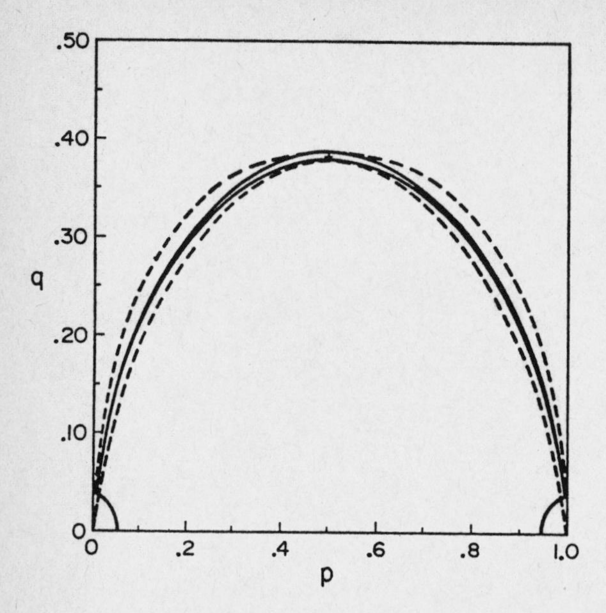

Figure 3 (a)

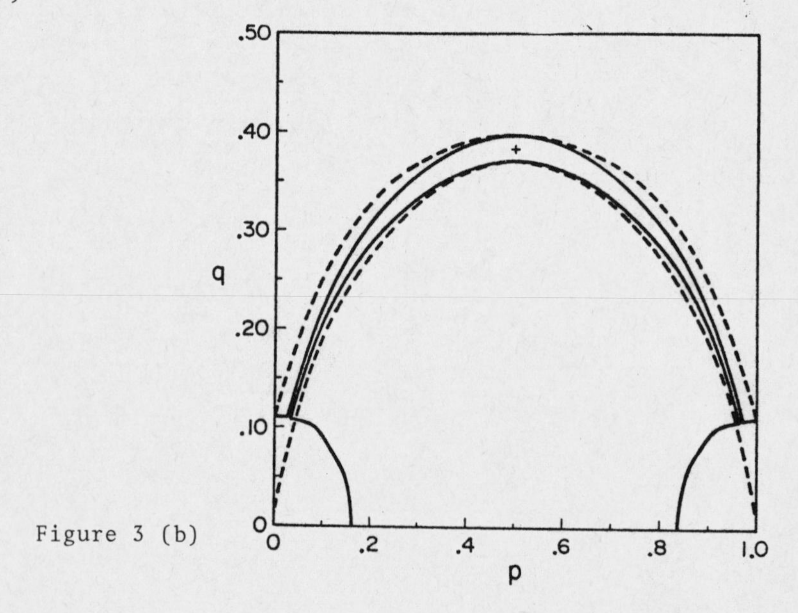

Figure 3 (b)

——— stability boundaries for capillary waves; ----- stationary disturbances. +, point of maximum growth rate. Region inside solid lines shows region of instability. (a) $h/\lambda = 0.03$, (b) $h/\lambda = 0.09$.

REFERENCES

[1] B. Chen and P. G. Saffman, Steady gravity-capillary waves on deep water I. Weakly nonlinear waves, Stud. Appl. Math. 60: 183-210 (1979).

[2] B. Chen and P. G. Saffman, Numerical evidence for the existence of new types of gravity waves of permanent form on deep water, Stud. Appl. Math. 62: 1-21 (1980).

[3] B. Chen and P. G. Saffman, Steady gravity-capillary waves on deep water II. Numerical results for finite amplitude, Stud. Appl. Math. 62: 95-111 (1980).

[4] G. D. Crapper, An exact solution for progressive capillary waves of arbitrary amplitude, J. Fluid Mech. 2: 532-540 (1957).

[5] M. S. Longuet-Higgins, The instabilities of gravity waves of finite amplitude in deep water I. Superharmonics, Proc. Roy. Soc. Lond., A 360: 471-488 (1978).

[6] M. S. Longuet-Higgins, The instabilities of gravity waves of finite amplitude in deep water II. Subharmonics, Proc. Roy. Soc. Lond., A 360: 489-505 (1978).

[7] M. S. Longuet-Higgins and M. J. H. Fox, Theory of the almost-highest wave: the inner solutions, J. Fluid Mech. 80: 721-741 (1977).

[8] J. W. McLean, Instabilities of finite-amplitude water waves, J. Fluid Mech. 114: 315-330 (1982).

[9] J. W. McLean, Y. C. Ma, D. U. Martin. P. G. Saffman and H. C. Yuen, Three-dimensional instability of finite amplitude water waves, Phys. Rev. Lett. 46: 817-820 (1981).

A BLOCK 5(4) EXPLICIT RUNGE-KUTTA FORMULA WITH "FREE" INTERPOLATION.

J.R. CASH

Department of Mathematics, Imperial College, South Kensington, London SW7, ENGLAND.

ABSTRACT

Embedded pairs of explicit Runge-Kutta formulae have been widely used for the numerical integration of non-stiff systems of first order ordinary differential equations. Because explicit Runge-Kutta formulae do not in general have a natural underlying interpolant, most implementations of these formulae restrict the steplength of integration so as to "hit" all output points exactly. Clearly this will normally lead to gross inefficiency when output is requested at many points and this is widely recognised as being a major disadvantage of explicit Runge-Kutta formulae. In addition there are some classes of problems for which an interpolation capability is indispensible. Recently the present author has proposed the use of block explicit Runge-Kutta formulae which advance the integration by more than one step at a time. One of the advantages of these block formulae is that they require less function evaluations per step than standard explicit Runge-Kutta formulae of the same order. In this paper we analyse completely the 5(4) two step Runge-Kutta formula with the minimum number of stages and show that it is possible to obtain a Runge-Kutta formula of this class with "free" interpolation capabilities. Some numerical results are given to compare the performance of a particular block 5(4) Runge-Kutta method with that of the widely used code RKF45 of Shampine and Watts.

1. INTRODUCTION

In a recent paper [11], Shampine has highlighted the difficulty of efficiently obtaining accurate numerical solutions at intermediate ("off-step") points when using explicit Runge-Kutta formulae for the integration of the non-stiff initial value problem

$$\frac{dy}{dx} = f(x,y), \quad y(x_0) = y_0, \quad y \in \mathbb{R}^t. \tag{1.1}$$

The widely used backward differentiation formulae for stiff equations and Adams formulae for non-stiff equations both have a natural underlying interpolating polynomial (see e.g. 10 p111) (in fact Adams formulae have several such polynomials) and so solutions at off-step points can be obtained simply by evaluating this polynomial at an appropriate point. This in turn means that the steplengths of integration used with these formulae are determined by accuracy requirements alone and are practically independent of the distribution of output points. (Note however that although this interpolation capability has been widely used with BDF and Adams methods, the theory supporting it in a variable step/variable order mode is far from complete [11]). In contrast, explicit Runge-Kutta formulae do not normally have such an interpolating polynomial available and this means that most implementations of explicit Runge-Kutta formulae [9,13] choose the stepsize of integration so as to hit all output points exactly. This can, of course, lead to an extremely expensive integration if there are many output points [3] and this generally makes Runge-Kutta formulae uncompetitive with Adams formulae in this situation. One obvious example of this is when graphical output is required and in this case it is normally recommended that Runge-Kutta formulae should not be used. In addition there are certain inverse problems associated with (1.1), such as that of finding x where some function

$$g(x,y(x), y'(x)) = 0,$$

where an interpolation capability is practically indispensible.

An examination of the interpolation problem for fourth and fifth order explicit Runge-Kutta formulae has been carried out by Horn [5,6,7]. Horn derives fourth and fifth order formulae which, when used with a steplength h in the interval $[x_n, x_{n+1}]$ can compute fourth or fifth order solutions at any point $x_n + \sigma h$, $0 \leqslant \sigma \leqslant 1$. Although this approach seems to be promising, it can be regarded as being rather expensive since the interpolation capability for fourth order formulae requires one extra function evaluation per step while for the fifth order formula five extra function evaluations are required per step. In addition the interpolants have the possible drawback that they are not c^1 continuous.

In this paper we consider a different solution to this problem through the use of block Runge-Kutta formulae. A standard explicit Runge-Kutta formula for the integration of (1.1) can be written in the form

$$y_{n+1} - y_n = h \sum_{i=1}^{q} b_i k_i \tag{1.2a}$$

$$k_i = f(x_n + c_i h, y_n + h \sum_{j=1}^{i-1} a_{ij} k_j), \quad 1 \leqslant i \leqslant q. \tag{1.2b}$$

The characterising properties of such formulae are that they are one step in nature and obtain a high order of accuracy by means of repeated function evaluations. Two widely used implementations of explicit Runge-Kutta formulae, which can be regarded as representing the state of the art, are the codes RKF45 of Shampine and Watts [13], which implements a 5(4) pair, and DVERK of Hull et. al[9] which implements a 6(5) pair. As an alternative to (1.2a,b), the present author has proposed the use of block explicit Runge-Kutta formulae. These have been defined in [3] as having the general form

$$y_{n+i} - y_n = h \sum_{j=1}^{s} b_{ij} k_j, \quad \sum_{j=1}^{s} b_{ij} = i, \quad 1 \leqslant i \leqslant \nu \tag{1.3a}$$

$$k_j = f(x_n + c_j h, \; y_n + h \sum_{i=1}^{j-1} a_{ji} k_i), \quad 1 \leqslant j \leqslant s. \tag{1.3b}$$

A formula of this type advances ν steps using a total of s function evaluations and so it is natural to speak of it as requiring s/ν functions per step. In [2] several low order formulae of this type were presented and in particular it was shown that it is possible to derive a fourth order block formula requiring only 3 function evaluations per step. Extensive numerical results presented in [3] indicate that a particular block 5(4) explicit Runge-Kutta formula with $s=9, \nu=2$ is very competitive with RKF45 particularly when stringent tolerances are imposed. However all formulae considered in [2,3] have the possible drawback that they require one extra function evaluation per step in order to perform interpolation. In addition all coefficients of the 5(4) formula were given as the result of a numerical search i.e. not in rational form. The purpose of the present paper is to give a complete analysis of the 5(4) case with $\nu=2$, $s=9$ and to derive a class of formulae for which interpolation is free in the sense that no extra function evaluations are required. A particular formula of this class is derived, and analysed in detail, and its performance is compared with that of RKF45 on a large class of test problems.

2. THE FORMULAE

The formulae to be analysed in this section are of the form (1.3a,b) with $s=9$, $\nu=2$. We consider the case $s=9$ since it was shown in [3] that, in certain circumstances, this is the minimum possible number of function evaluations needed to obtain a block 5(4) formula with two steps. We restrict ourselves to the case $\nu=2$ since we feel that it is better to understand fully the computational aspects of two step blocks before considering larger blocks. However we expect that the case $\nu>2$ will prove worthy of investigation particularly for use with strict tolerances.

In order to compute the solution y_{n+1} at the first step of the block starting from x_n we use a six stage formula of the form

$$y_{n+1} - y_n = h \sum_{j=1}^{6} b_{1j} k_j. \tag{2.1}$$

In [3] we imposed the restriction that this formula should be the standard 5(4) pair used by Shampine and Watts in RKF45. The main reason for adopting this approach is that we had originally hoped to implement (1.3a,b) so that the value of ν is chosen automatically - and this is indeed still a long term aim. This seems desirable since, although our numerical results show that the two step block is clearly more efficient in general than the one step scheme RKF45, it is equally clear that there are cases where the one step scheme is to be preferred. If we are to exploit the possibility of having variable ν it is important that the formula used to integrate the first step is a thoroughly tested and efficient one. However the investigation of efficient Runge-Kutta methods is still a very active area (see e.g. [12]) and it may be that RKF45 will be superceded as the most widely used 5(4) Runge-Kutta code in the future. In view of this it is appropriate to give our block 5(4) formula in some generality rather than using a special choice for (2.1). The general 5(4) formula is based on (2.1) where the fifth order solution is computed using (2.1), the embedded fourth order solution is computed using

$$\bar{y}_{n+1} - y_n = h \sum_{j=1}^{5} \bar{b}_{1j} k_j \tag{2.2}$$

and the k_j are defined by (1.2b). Formulae for these coefficients in terms of the c_i have been given by England [4] and are: For any c_2, c_3, c_4, c_5, c_6 which are distinct and non-zero such that

$$3 - 12c_3 + 10c_3^2 \neq 0$$

$$c_3 = 2c_4(1-4c_3+5c_3^2),$$

$$\bar{b}_{15} = 1/12[3 - 4(c_3+c_4) + 6c_3c_4]/[c_5(c_5-c_3)(c_5-c_4)]$$

$$\bar{b}_{14} = 1/12[3 - 4(c_3+c_5) + 6c_3c_5]/[c_4(c_4-c_5)(c_4-c_3)]$$

$$\bar{b}_{13} = 1/12[3 - 4(c_4+c_5) + 6c_4c_5]/[c_3(c_3-c_4)(c_3-c_5)]$$

$$\bar{b}_{12} = 0, \quad \bar{b}_{11} = 1 - (\bar{b}_{13}+\bar{b}_{14}+\bar{b}_{15}).$$

$$b_{16} = 1/60[12 - 15(c_3+c_4+c_5) + 20(c_4c_5+c_5c_3+c_3c_4)-30c_3c_4c_5]/[c_6(c_6-c_3)(c_6-c_4)(c_6-c_5)]$$

$$b_{15} = 1/60[12 - 15(c_3+c_4+c_6) + 20(c_3c_4+c_4c_6+c_6c_3)-30c_3c_4c_6]/[c_5(c_5-c_6)(c_5-c_3)(c_5-c_4)]$$

$$b_{14} = 1/60[12 - 15(c_3+c_5+c_6) + 20(c_3c_6+c_3c_5+c_5c_6)-30c_3c_5c_6]/[c_4(c_4-c_5)(c_4-c_6)(c_4-c_3)]$$

$$b_{13} = 1/60[12 - 15(c_4+c_5+c_6) + 20(c_5c_6+c_4c_6+c_4c_5)-30c_4c_5c_6]/[c_3(c_3-c_4)(c_3-c_5)(c_3-c_6)]$$

$$b_{12} = 0, \quad b_{11} = 1 - (b_{13}+b_{14}+b_{15}+b_{16}).$$

$$a_{32} = \tfrac{1}{2}c_3^2/c_2$$

$$a_{42} = a_{32}c_4(3-12c_4+10c_4^2)/[c_3(3-12c_3+10c_3^2)]$$

$$a_{52} = a_{32}c_5(3-12c_5+10c_5^2)/[c_3(3-12c_3+10c_3^2)]$$

$$a_{62} = a_{32}c_6(3-12c_6+10c_6^2)/[c_3(3-12c_3+10c_3^2)]$$

$$a_{43} = c_4(c_4-c_3)(c_3+c_4-4c_3c_4)/[c_3^2(3-12c_3+10c_3^2)]$$

$$a_{53} = c_5(c_5-c_3)[c_3+c_5-4c_3c_5-\tfrac{1}{2}c_4(3-10c_3c_5)]/[c_3(c_3-c_4)(3-12c_3+10c_3^2)]$$

$$a_{54} = \tfrac{1}{2}c_5(c_5-c_3)(c_3-2c_5(1-4c_3+5c_3^2))/[c_4(3-12c_3+10c_3^2)(c_3-c_4)]$$

$$a_{63}c_3 + a_{64}c_4 + a_{65}c_5 = \tfrac{1}{2}c_3c_6(c_6-c_3)(3-10c_3c_6)/[c_3(3-12c_3+10c_3^2)]$$

$$a_{63}c_3^2 + a_{64}c_4^2 + a_{65}c_5^2 = c_6(c_6-c_3)(c_3+c_6-4c_3c_6)/(3-12c_3+10c_3^2)$$

$$a_{63}c_3^3 + a_{64}c_4^3 + a_{65}c_5^3 = [1/20-[b_{14}a_{43}c_3^3+b_{15}(a_{53}c_3^3+a_{54}c_4^3)]]/b_{16}$$

$$a_{31} = c_3 - a_{32}$$

$$a_{41} = c_4 - (a_{42}+a_{43})$$

$$a_{51} = c_5 - (a_{52}+a_{53}+a_{54})$$

$$a_{61} = c_6 - (a_{62}+a_{63}+a_{64}+a_{65}).$$

The possibility of having a formula with free interpolation now comes from choosing

$$a_{7i} = b_{1i}, \quad 1 \leq i \leq 6. \tag{2.3}$$

This particular choice means that $k_7 \equiv y'_{n+1}$. If we now evaluate y'_{n+2} at the end of the block $[x_n, x_{n+2}]$ ready for use in the next step (y'_{n+2} is the k_1 for use in the next block) we have the data

$$y_n, y'_n, y_{n+1}, y'_{n+1}, y_{n+2}, y'_{n+2}$$

available in $[x_n, x_{n+2}]$. A quintic Hermite interpolating polynomial can be fitted to this data and, by evaluating this polynomial at the appropriate point, accurate solutions can be computed at any point in $[x_n, x_{n+2}]$. This interpolation procedure is free in the sense that no extra function evaluations are required to compute the "off-step" solution and clearly with this approach our step control procedure is based entirely on accuracy considerations and not on the positioning of the output points.

We now consider the solution of the order relations at n+2. We adopt the approach described in [3] with the main formula (1.3a,b) being of order 6 while the embedded formula

$$\bar{y}_{n+2} - y_n = h \sum_{j=1}^{9} b_{2j} k_j \tag{2.4}$$

is of order 4. We will re-scale the steplength from 2h to h so that we can make use of the extensive theory already available (see e.g. Butcher [1]). The solution procedure which we now adopt, which is a modification of that described in [3], is

1) Choose a_{7i} as in (2.3), set $c_7 = \frac{1}{2}$, $c_9 = 1$, choose c_8 arbitrarily.
2) Choose b_{26} arbitrarily, $b_{22} = 0$.
3) Set $A_1 = -1/20 + 1/12c_8 + (1/12 - 1/6c_8)c_7$

$A_2 = -1/30 + 1/20c_8 + (1/20 - 1/12c_8)c_7$

Then

$$b_{23} = (A_2 - c_4A_1 - b_{26}c_6(c_6-c_4)(c_6-1)(c_6-c_8)(c_6-c_7))/[c_3(c_3-c_4)(c_3-1)(c_3-c_8)(c_3-c_7)]$$

$$b_{24} = (A_1 - c_3(c_3-1)(c_3-c_8)(c_3-c_7)b_{23} - b_{26}c_6(c_6-1)(c_6-c_8)(c_6-c_7))/[c_4(c_4-1)(c_4-c_7)(c_4-c_8)]$$

$$b_{25} = -[(c_3-1)(c_3-c_8)b_{23}a_{32} + (c_4-1)(c_4-c_8)b_{24}a_{42} + b_{26}(c_6-1)(c_6-c_8)a_{62}]/[c_5(c_5-1)(c_5-c_8)a_{52}]$$

$$b_{27} = [-1/12 + 1/6c_8 - c_7(c_3-1)(c_3-c_8)b_{23} - c_4(c_4-1)(c_4-c_8)b_{24} - c_5(c_5-1)(C_5-c_8)b_{25} - c_6(c_6-1) \times (c_6-c_8)b_{26}]/[c_7(C_7-1)(c_7-c_8)]$$

$$b_{28} = (-1/6 - b_{23}c_3(c_3-1) - b_{24}c_4(c_4-1) - b_{25}c_5(c_5-1) - b_{27}c_7(c_7-1) - b_{26}c_6(c_6-1))/[c_8(c_8-1)]$$

$$b_{29} = \tfrac{1}{2} - b_{23}c_3 - b_{24}c_4 - b_{25}c_5 - b_{26}c_6 - b_{27}c_7 - b_{28}c_8$$

$$b_{19} = 1 - b_{23} - b_{24} - b_{25} - b_{26} - b_{27} - b_{28} - b_{29}.$$

4) Set $Y1 = \sum_{i=1}^{6} b_i(c_i-1)a_{i2}$, $Y2 = \sum_{i=1}^{6} b_i(c_i^2-1)a_{i2}$

Then

$$a_{82} = -(Y1 + b_{27}(c_7-1)a_{72})/[b_{28}(c_8-1)]$$

5) Set $Z1 = -1/60 - \sum_{i=1}^{6} b_{2i}(c_i-1)a_{ij}c_j^2$, $Z2 = -1/36 - \sum_{i=1}^{6} b_{2i}(c_i^2-1)a_{ij}c_j^2$

Then $\Sigma a_{7j}c_j^2 = (Z1(c_8+1)-Z2)/[(c_7-1)(c_8-c_7)b_{27}] \equiv AIJCJ(7)$ say

$$\Sigma a_{8j}c_j^2 = (Z1-b_{27}(c_7-1)\Sigma a_{7j}c_j^2)/[b_{28}(c_8-1)] \equiv AIJCJ(8).$$

6) Choose a_{83} arbitrarily and solve the following linear system for a_{84}, a_{85}, a_{86}, a_{87}:

$$\sum_i a_{8i}c_i = \tfrac{1}{2}c_8^2$$

$$\sum_i a_{8i}c_i^2 = AIJCJ(8)$$

$$\sum_i a_{8i}c_i^3 = -[1/120 + \sum_i b_{2i}(c_i-1)a_{ij}c_j^3]/[b_{28}(c_8-1)]$$

$$-\sum_i b_{2i}(c_i-1)a_{ij}a_{j2} + \sum_i b_{2i}(c_i-1)a_{i2} + b_{28}(c_8-1)a_{82} = \sum_i b_{28}(c_8-1)a_{8i}a_{i2}.$$

7) Compute a_{81} from

$$a_{81} = c_8 - \sum_{i=2}^{7} a_{8i}$$

8) Compute a_{9i}, $i=1,2,\ldots,8$ from

$$a_{9i} = [b_{2i}(1-c_i) - \sum_{j=i+1}^{8} b_{2j}a_{ji}]/b_{29}.$$

These eight steps give all the coefficients for the main integration formula.

To find the coefficients for the embedded formula:

9) Choose $\bar{b}_{27}$, $\bar{b}_{28}$, $\bar{b}_{29}$ arbitrarily. Compute $\bar{b}_{23},\bar{b}_{24},\bar{b}_{25},\bar{b}_{26}$ as the solution of the four linear equations

$$\bar{b}_{23}c_3+\bar{b}_{24}c_4+\bar{b}_{25}c_5+\bar{b}_{26}c_6 = \tfrac{1}{2}-\bar{b}_{27}c_7-\bar{b}_{28}c_8-\bar{b}_{29}c_9$$

$$\bar{b}_{23}c_3^2+\bar{b}_{24}c_4^2+\bar{b}_{25}c_5^2+\bar{b}_{26}c_6^2 = 1/3-\bar{b}_{27}c_7^2-\bar{b}_{28}c_8^2-\bar{b}_{29}c_9^2$$

$$\bar{b}_{23}c_3^3+\bar{b}_{24}c_4^3+\bar{b}_{25}c_5^3+\bar{b}_{26}c_6^3 = \tfrac{1}{4}-\bar{b}_{27}c_7^3-\bar{b}_{28}c_8^3-\bar{b}_{29}c_9^3$$

$$\bar{b}_{23}a_{32}+\bar{b}_{24}a_{42}+\bar{b}_{25}a_{52}+\bar{b}_{26}a_{62} = -\bar{b}_{27}a_{72}-\bar{b}_{28}a_{82}-\bar{b}_{29}a_{92}.$$

10) Compute $\bar{b}_{21}$ from $\bar{b}_{21} = 1-\sum_{i=3}^{9}\bar{b}_{2i}$

The above gives a simple algorithm for computing all of the coefficients of our block 5(4) formula in terms of the arbitrary coefficients c_2, c_3, c_5, c_6, c_8, b_{26}, a_{83}. We now discuss the way in which these free parameters are to be chosen. As explained earlier we choose c_2, c_3, c_5 c_6 so that the formula used to integrate from x_n to x_{n+1} is the Fehlberg pair used in RKF45. We do this because this particular formula is well known and has been very widely tested. However we emphasise that the analysis we have given is for a general choice of the c_i and a block formula can easily be derived using any desired formula to compute y_{n+1}. We choose $c_8 = 3/4$ so as to equi-distribute the c_i in $[x_{n+1},x_{n+2}]$. This leaves b_{26} and a_{83} to be chosen and, as we will now explain, we will specify these coefficients so as to satisfy certain accuracy and stability requirements.

The local truncation error associated with the main method (i.e. the fifth order method) at n+1 is of the form

$$LTE^{(M)} = h^6 \sum_{i=1}^{N_6} T_i^{(6)} D_i^{(6)} + O(h^7) \tag{2.5}$$

where the $\{T_i^{(6)}\}$ are numerical constants depending on the parameters of the method and the $\{D_i^{(6)}\}$ are elementary differentials of order 6 (see Butcher [1]). Similarly the embedded formula at n+1 has a local truncation error of the form

$$LTE^{(E)} = h^5 \sum_{i=1}^{N_5} T_i^{(5)} D_i^{(5)} + O(h^6). \tag{2.6}$$

At the point n+2 the local truncation error of the main formula would generally have the form

$$\overline{LTE}^{(M)} = (2h)^6 \sum_{i=1}^{N_6} \bar{T}_i^{(6)} D_i^{(6)} + O(h^7) \tag{2.7}$$

while the local truncation error of the embedded formula is of the form

$$\overline{LTE}^{(E)} = (2h)^5 \sum_{i=1}^{N_5} \bar{T}_i^{(5)} D_i^{(5)} + O(h^6). \tag{2.8}$$

Now if a conventional Runge-Kutta method was to be used with a fixed stepsize h to integrate two steps from x_n to x_{n+2} we would expect asymptotically (for small h) that the local error in integrating from x_n to x_{n+1} is the same as that committed in integrating from x_{n+1} to x_{n+2}. We could mirror this behaviour with block Runge-Kutta methods by constraining our parameters so that

$$\overline{LTE}^{(M)} = 2 \times LTE^{(M)}, \quad \overline{LTE}^{(E)} = 2 \times LTE^{(E)}.$$

In [2] this was called the equi-distribution property. However what we will actually do, for reasons which we will explain, is to make

$$|\bar{T}_i^{(6)}| \leq \frac{1}{2^5} |T_i^{(6)}| \text{ and } |\bar{T}_i^{(5)}| \leq \frac{1}{2^4} |T_i^{(5)}| \text{ for all i.} \tag{2.9}$$

For the main formula we can achieve this aim by choosing $\bar{T}_i^{(6)} \equiv 0$ for all i. We do this so that, after performing local extrapolation, we actually carry forward a sixth order solution at the end of the block. Our numerical experience indicates that this normally gives a much smaller global error than is obtained with RKF45. Although we do carry forward a sixth order solution it would be wrong to regard our method as being a sixth order one since the error control is in a fourth order solution. This is in contrast to standard sixth order Runge-Kutta methods, such as DVERK, which control the error in an embedded fifth order solution. The reason why we allow an inequality rather than demanding equality in (2.9) is on account of our step control procedure and this we will now explain. The procedure used to control h is as follows:

1) Starting from x_n compute 5th and 4th order solutions y_{n+1}, $\bar{y}_{n+1}$ at n+1 using (2.1), (2.2). Estimate the local truncation error in the lower order solution $\bar{y}_{n+1}$ by

$$LTE_{n+1} = y_{n+1} - \bar{y}_{n+1}.$$

If $\|LTE_{n+1}\| > Tol$ decrease h accordingly. If however $\|LTE_{n+1}\| < Tol$ then compute y_{n+2}, $\bar{y}_{n+2}$ at n+2 using (1.3a), (2.4) and estimate the local error, LTE_{n+2}, in $\bar{y}_{n+2}$ using

$$LTE_{n+2} = y_{n+2} - \bar{y}_{n+2}$$

Now in choosing the free parameters it is important to ensure that as many values of $T_i^{(5)}$ and $\bar{T}_i^{(5)}$, $i = 1,2,\ldots,N_5$, as possible are non-zero to give an effective error control. In the formulae we shall give, we have made all of these constants non-zero. However for some problems there may be an unfortunate cancellation of terms in the sums

$$\Sigma T_i^{(5)} D_i^{(5)} \text{ or } \Sigma \bar{T}_i^{(5)} D_i^{(5)}$$

so making the error control ineffective at these points. If however we insisted that $T_i^{(5)} = 2^4 \bar{T}_i^{(5)}$ for all i the same cancellation of terms could occur at both n+1 and n+2, particularly if the elementary differentials $D_i^{(5)}$ are slowly varying at these points, and we may have the problem of accepting a solution when we should not do so. In the formulae given in Table 1 we have allowed inequality in (2.9) and for these formulae the ratios $T_i^{(5)}/\bar{T}_i^{(5)}$ are different for all i. This in turn means that the problem of cancellation among the constituents of (2.5) which can be a serious problem with conventional Runge-Kutta formulae, is more likely to be detected and dealt with by block formulae and this hueristic argument is reflected in our numerical results where we find that the block formulae are often more reliable than single step formulae.

The final consideration is that of stability. Since our method is basically two step in nature we would like to have an interval of absolute stability which is at least twice as large as that of the conventional formula used to integrate from n to n+1. It was found that this requirement together with (2.9), was satisfied for a range of the free parameters b_{26}, a_{83}. One particular such formula is given in Table 1 where for the sake of simplicity we have given the coefficients of this formula scaled on an interval h. The regions of absolute stability of both the main formula and the embedded formula scaled on an interval 2h are given in Figure 1. One of our considerations in deriving these formulae was to obtain an embedded formula and a main formula with comparable regions of absolute stability. It can be seen from Figure 1 that we have been quite successful in this aim and in particular the intervals of absolute stability of the main and embedded formulae are approximately (-7.3,0) and (-7,0) respectively.

Finally we present some numerical results illustrating the performance of a variable step code based on the block formula presented in Table 1. Given the existence of the high quality code RKF45, it is not difficult to write a reasonable general purpose code based on our block formula since the modifications to RKF45 needed to achieve this are not drastic. Perhaps the only major conceptual change from RKF45 concerns the choice of steplength after a successful step (block). Suppose we are currently using a steplength h, an absolute accuracy of ε is requested and an estimate of the local error at the present point is ESTTOL. Then, assuming the current step is successful, the new step h_{NEW} is chosen so that

$$h_{NEW} = SF \times h \times (\varepsilon/ESTTOL)^{1/5} \qquad (2.10)$$

where SF is a "safety factor" taken to be 0.9 in RKF45. In our block code we have

$$SF = SF(\varepsilon).$$

0									
1/8	1/8								
3/16	3/64	9/64							
6/13	966/2197	-3600/2197	3648/2197						
1/2	439/432	-4	1840/513	-845/8208					
1/4	-4/27	1	-1772/2565	1859/8208	-11/80				
	8/135	0	3328/12825	28561/112860	-9/100	1/55			
	25/432	0	704/2565	2197/8208	-1/10	0			
1/2	8/135	0	3328/12825	28561/112860	-9/100	1/55			
3/4	-175733/447400	39195/29081	-13468988/69067375	-302656523/187013200	3173949/4474000	-453213/1230350	2835/2237		
1	2807770163/2023015500	-552/239	-620146592/229888125	19748597921/2023015500	-38023362/12099375	11122228/2957625	-31616/4541	2210156/1839105	
	229957/3201120	0	157696/1500525	371293/1231200	1/100	1/5	-2/15	42503/115425	4541/59280
	2934384339/47081088000	0	46046869/229888125	5014065901/26156160000	16096591/726560000	1/10	0	265234379/774360000	81/1000

Table 1

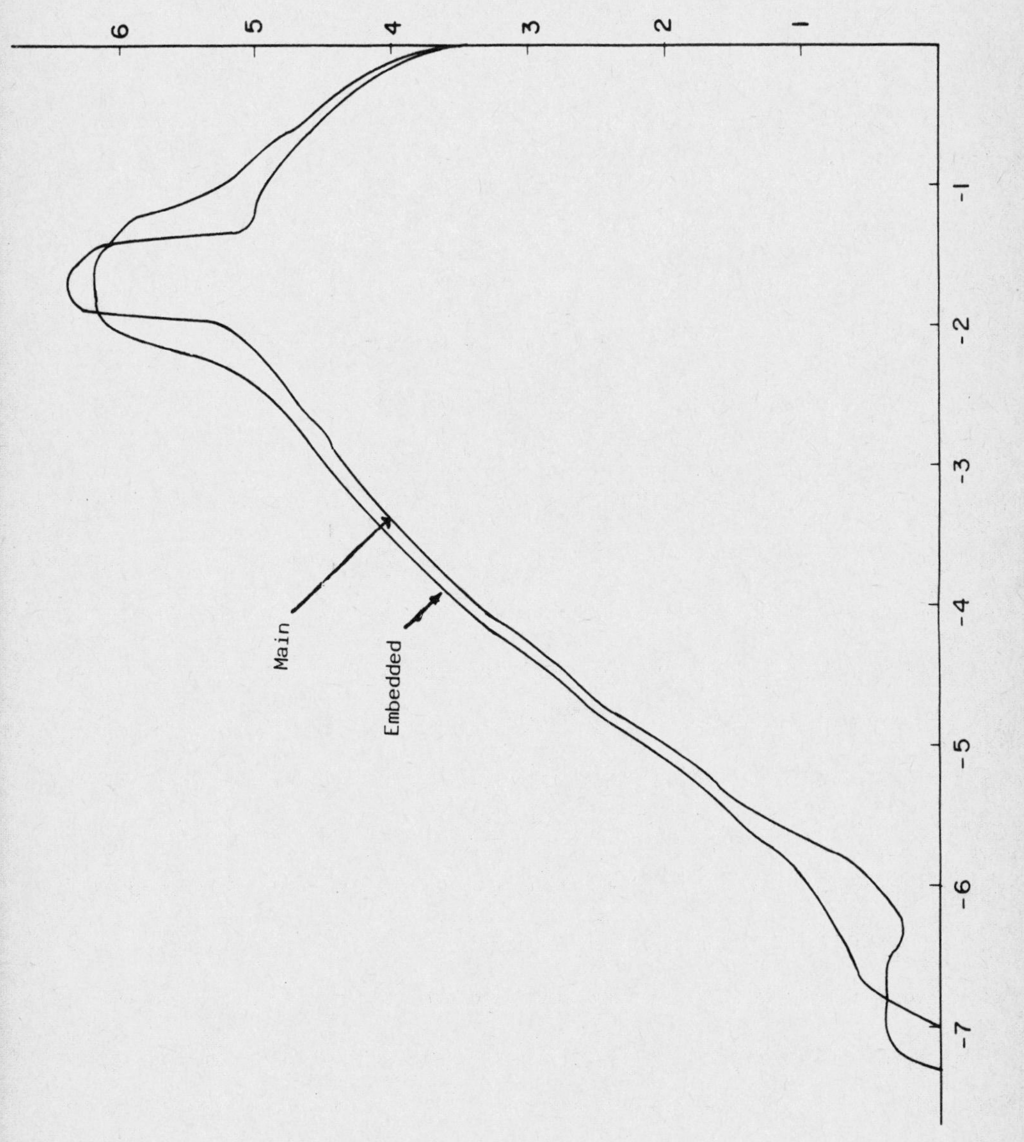

FIGURE .1.

In practice we use relation (2.10) to choose the new step with

$$SF(\varepsilon) = \begin{cases} 0.7 & \text{if} \quad \varepsilon > 10^{-4} \\ 0.8 & \text{if} \quad 10^{-4} \geq \varepsilon > 10^{-6} \\ 0.95 & \text{if} \quad \varepsilon \leq 10^{-6} \end{cases}$$

The actual choice of the (peicewise constant) scaling factor $SF(\varepsilon)$ does not seem to be crucial but the idea of using a small SF for crude tolerances and one near to unity for very strict tolerances does seem to be a useful one although, of course, we do not claim this idea to be new. Indeed we would expect SF to vary with ε since the smaller the value of h the more valid our asymptotic analysis should be and consequently we are safer in taking SF close to 1. The converse argument holds for crude tolerances. In Table 2 we give the numerical results obtained for the well known test set of Hull et. al [8]. We cannot claim that this validates our algorithm because we feel that the purpose of a test set should be to eliminate poor methods. However producing results on a standard test set does serve at least two useful purposes, firstly it allows comparison with other methods and secondly by running our method on a large set of problems(25 test problems at 7 tolerances i.e. 175 problems) we can have some confidence in the approach if our theoretical analysis is reflected in the numerical results. In Table 2 we compare the results obtained using the block formula of Table 1 and the code RKF45 on the test set using pure absolute error tolerances $10^{-3}, 10^{-4},\ldots,10^{-9}$. In Table 2 we give the number of function evaluations, the number of steps deceived (i.e. the number of times an accepted solution at any point had a local error exceeding the tolerance) and the maximum local error in units of the tolerance. It can be seen from the results given in Table 2 that on this test set the block formula is more efficient than RKF45 (by a factor of about 14%) is about twice as reliable and produces a smaller "Maximum Local Error". It should also be remembered that the block formula allows free interpolation and has the advantage of carrying forward a sixth order solution.

Finally we should point out that our formulae can be regarded simply as standard Runge-Kutta methods. The special properties which they have is that they require significantly more than the minimum number of stages to achieve a specified order, they have enlarged regions of absolute stability and the free coefficients are chosen so as to give small error constants at the end of the integration step and a solution with the desired asymptotic accuracy at the middle of the step (together with embedded local error estimates at both these points). In addition the computation is arranged so that, if the required accuracy at the mid-step point is not achieved, the solution at the end of the step is not computed but the step is instead reduced. In addition the local error estimate for any step is taken to be the maximum of that at the middle of the step and at the end of the step. The investigation of these methods is at present at an early stage. But however these methods are regarded, either as block methods or standard Runge-Kutta methods, we feel that the results which we have presented indicate that they are worth considering as possible alternatives to the standard Runge-

Kutta methods currently in use.

Tolerance	Block Formula F^n Evals	Steps Deceived	Max. Local Error	RKF45 F^n Evals	Steps Deceived	Max. Local Error
10^{-3}	4773	42	3.4	4858	114	9.3
10^{-4}	6493	28	4.8	6754	69	3.4
10^{-5}	8991	31	5.6	9514	20	1.4
10^{-6}	12197	12	3.3	13737	14	6.9
10^{-7}	17238	3	1.9	19842	6	2.1
10^{-8}	25967	2	2.0	29072	4	1.5
10^{-9}	37019	0	0.9	44946	2	1.3
overall	112678	118	5.6	128723	227	9.3

Table 2

Relative performance of block formula and RKF45 on 25 test problems.

References

1. J.C. Butcher, Coefficients for the study of Runge-Kutta integration processes, J. Austral.Math.Soc., 3, 1963, pp185-201.
2. J.R. Cash, Block Runge-Kutta methods for the numerical integration of initial value problems in ordinary differential equations, Part 1 - the non-stiff case, Math. Comp. 40, 1983, pp175-192.
3. J.R. Cash, Block embedded explicit Runge-Kutta methods, J. Comp. and Math. with Applics., to appear.
4. R. England, Error estimates for Runge-Kutta type solutions to systems of ordinary differential equations, Computer J., 12, 1969, pp166-170.
5. M.K. Horn, Scaled Runge-Kutta algorithms for handling dense output,Rep. DFVLR-FB81-13, DFVLR, Oberpfaffenhofen, F.R.G, 1981.
6. M.K. Horn, Scaled Runge-Kutta algorithms for treating the problem of dense output Rep NASA TMX-58239, L.B. Johnson Space Center, Houston, Tx., 1982.
7. M.K. Horn, Fourth-and fifth order, scaled Runge-Kutta algorithms for treating dense output, SIAM J. Numer.Anal. 20, 1983, pp558-568.
8. T.E. Hull, W.H. Enright, B.M. Fellen and A.E. Sedgewick, Comparing numerical methods for ordinary differential equations, SIAM J.Numer.Anal.,9, 1972, pp603-637.
9. T.E. Hull, W.H. Enright and K.R. Jackson, User's guide for DVERK - a subroutine for solving non-stiff ODE's, Rep 100, Dept. Computer Science, University of Toronto, Canada, 1976.
10. J.D. Lambert, Computational Methods in Ordinary Differential Equations, London, Wiley 1973.

11. L.F. Shampine, Interpolation for Runge-Kutta methods, Rep SAND83-25 60, Sandia National Laboratories, January 1984

12. L.F. Shampine, Some practical Runge-Kutta formulas, Rep. SAND84-0812, Sandia National Laboratories, April 1984.

13. L.F. Shampine and H.A. Watts, DEPAC - design of a user oriented package of ODE solvers, Rep SAND 79-2374, Sandia National Laboratories, 1980.

SEQUENTIAL STEP CONTROL FOR INTEGRATION OF TWO-POINT BOUNDARY VALUE PROBLEMS

Roland England
IIMAS-UNAM
Apdo. Postal 20-726
01000 México, D.F.
México

Robert M.M. Mattheij
Mathematisch Instituut
Katholieke Universiteit
6525 ED Nijmegen
The Netherlands

1. Introduction

Many two-point boundary value problems have sharp boundary layers, or rapidly varying fundamental modes. A uniform discretization may then be inefficient, while an iterative adaptive refinement process may be undesirable, particularly for linear problems. A multiple shooting approach should permit step sizes to be chosen sequentially, based on the behaviour of a smooth particular solution, and numerically estimated layers. However, normal step control procedures, for initial value problem integrators, are based on asymptotic error estimates for small step sizes. These will normally be large in the presence of fast growing fundamental modes, unless unnecessarily small step sizes are used. In this paper, a special form of error indicator is described, which should permit the use of step sizes appropriate to the particular solution, for an appreciable number of integration steps, even in the presence of much faster growing fundamental modes. Except during a starting up stage, such an indicator is obtained as the difference between a special explicit predicted value and a matched implicit corrected value. The linked corrector formulae must be solved by a modified Newton iteration (directly for a linear problem) to obtain the desired stability properties. Efficient and stable methods for solving the resulting linear algebraic systems are discussed. Some preliminary results are also presented, to show the feasibility of finding smooth solutions with appropriate step sizes, by the use of suitable sets of predictor-corrector formulae.

2. Boundary Value Problems and Discretization

The first author was one of the pioneers to work on a general purpose program for solving two-point boundary value problems. The resulting multiple shooting code [4] implements strategies which attempt to choose appropriate shooting intervals, as well as step sizes for the basic discretization to control the error in each interval [3]. As a library routine, it has been successfully used on many problems.

A few of them are described in [15], [5]. However, for some problems, it is inefficient, or works better for a user who judiciously overrides some of the automatic options.

A better theory is now growing up [2], [14], [11], [6], and should improve the basis for selecting discretization formulae [7], and shooting intervals. This paper concerns the step size control for the basic discretization.

Consider the system of n differential equations:

$$dY/dt = \dot{Y} = G(Y) \in \mathbb{R}^n \quad (a \leq t \leq b) \tag{1}$$

with n boundary conditions

$$H_0 Y(a) + H_1 Y(b) = C \in \mathbb{R}^n \tag{2}$$

where H_0, H_1 are n×n matrices. It will be assumed that the problem is well posed, as discussed, for example, by [11]. The solution may have sharp boundary layers, and even internal layers, if $|\lambda_i|(b-a) >> 1$ for some of the eigenvalues λ_i of the Jacobian matrix $\partial G/\partial Y$.

A typical example is given by the equation:

$$\ddot{y} = \lambda^2 y + f(t) \tag{3}$$

where f(t) is a slowly varying function. For large λ, the solution in much of the interval is approximately $y \sim - f(t)/\lambda^2$, but the boundary conditions cause a singular perturbation (fig.1) which, to a level of significance TOL, extends a distance of approximately $\delta \sim -\ln(TOL)/\lambda$.

Defining

$$y_1 = \dot{y} + \lambda y \, , \quad y_2 = \dot{y} - \lambda y \tag{4}$$

the standard form (1) for the equation is:

$$\dot{y}_1 = \lambda y_1 + f(t) \, , \; \dot{y}_2 = -\lambda y_2 + f(t) \tag{5}$$

two independent equations, each with a boundary layer at one end. Normal initial value integrators will need a step size $h = O(1/\lambda)$ throughout (a,b), and to retain some accuracy in each shooting interval, the interval size Δt must not exceed $-\ln(\varepsilon)/\lambda$ where ε is greater than the machine accuracy. Thus both h and Δt must be uniformly small in (a,b), even where the solution varies very slowly. Such methods will use excessive computer time (determined by h) and storage (determined by Δt).

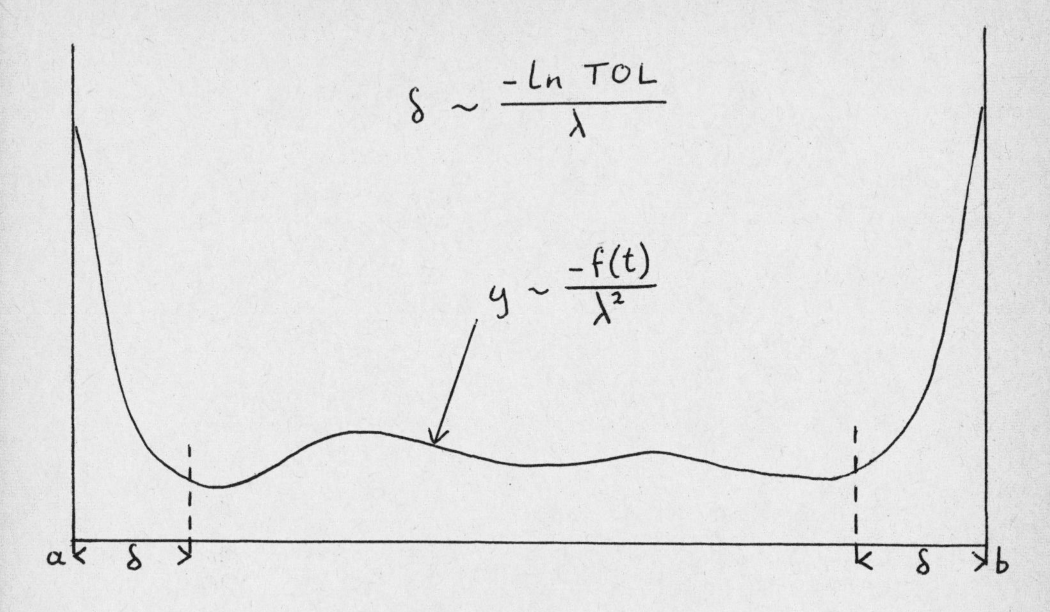

Figure 1

3. Adaptive Step Selection

In the global finite difference [13], and piecewise collocation [1] approaches, an initial discretization may well be of this uniform type, with iterative adaptive refinement to increase the accuracy where the solution varies rapidly. Sometimes such an approach also gives excessive refinement of the mesh where the solution is smooth. In any case, it uses a large amount of storage for approximations at the mesh points, and requires the solution of a number of discrete problems, simply to determine the mesh, which is particularly inefficient for linear problems.

A multiple shooting approach should permit step sizes to be chosen sequentially, fine in the boundary layers, and coarse in the smooth regions. However, special integration processes are required. For a linear problem, if a reasonably accurate particular solution can be found on the first integration, and the fundamental (complementary) modes are correspondingly accurate in the layers, the final solution can be found by simple superposition, since the fundamental modes are essentially zero outside the layers.

It is also important for the stability (or conditioning) of the discrete problem that fast decaying modes with $Re(\lambda) \ll -1/(b-a)$, which are controlled by initial conditions in the continuous problem [14], should be approximated by decaying numerical sequences equally controlled by initial conditions [6]. In the same way, fast growing modes with $\mathrm{Re}(\lambda) \gg 1/(b-a)$, which are controlled by terminal conditions, should be approximated by growing numerical sequences. In [6] this property is called dichotomic stability.

The new approach will need to identify the position of potential sharp layers, and estimate their width as discussed in [8]. Any normal step control procedure should select appropriately fine step sizes in these layers, but special action will be required to increase the step size outside the layers, a special discretization formula to maintain stability and avoid the growth of unwanted fast modes, and a special step control procedure to maintain the large step size while the particular solution remains smooth. This paper recalls the dichotomically stable formulae introduced in [7], and presents suitable procedures for their implementation and step control.

4. Dichotomically Stable Formulae

A k-step general linear scheme, when applied to the test equation $\dot{y}=\lambda y$ with constant step size h, gives rise to a recurrence relation:

$$Q_0(h\lambda)y_{i+1}+Q_1(h\lambda)y_i+\ldots+Q_k(h\lambda)y_{i-k+1}=0 \tag{6}$$

where the $Q_j(z)$ are polynomials in z. The general solution takes the form:

$$y_i = \Sigma_{j=1}^{k} c_j R_j(h\lambda)^i \tag{7}$$

where $R_j(z)$, $j=1,2,\ldots,k$, are the roots of the characteristic polynomial:

$$Q(z,R)=Q_0(z)R^k+Q_1(z)R^{k-1}+\ldots+Q_k(z) \tag{8}$$

and the c_j are arbitrary constants.

Dichotomic stability for real values of λ, implies that $|R_1(z)|<1$ for z (real) <0 and also $|R_1(z)|>1$ for $z>0$. For consistency, it is also necessary that $R_1(0)=1$. With the stability conditions indicated in [7], the other $R_j(z)$ must be less than unity in absolute value, and so $R_1(z)$ must remain real for all positive values of z. If it remains bounded, then it cannot change sign, and the limit $R_1(\infty)=1$, while $Q_0(z)$ has only complex conjugate roots.

For convenience, consider a characteristic polynomial quadratic in z:

$$Q(z,R)= -\Sigma_{j=0}^{k}(\alpha_j+\beta_j z+\gamma_j z^2)R^{k-j} \tag{9}$$

and let the spurious roots $R_j(z)$, $j=2,3,\dots,k$ satisfy $R_j(0)=R_j(\infty)=0$ to give large regions of dichotomic stability. Then $\alpha_j=\gamma_j=0$, $j=2,3,\dots,k$, while $\alpha_1=-\alpha_0$, $\gamma_1=-\gamma_0$. Normalizing $\alpha_0=-1$ leaves

$$Q(z,R)=R^k-R^{k-1}-z\,\Sigma_{j=0}^{k}\beta_j R^{k-j}-z^2\gamma(R^k-R^{k-1}) \tag{10}$$

where the coefficients β_j $(j=0,1,\dots,k)$ and γ may be chosen [7] to give a scheme of order $p=k+2$, meaning that $R_1(z)-e^z=O(z^{p+1})$ as $z\to 0$. For $p\leqslant 11$, these schemes are $A(\alpha)$-stable, and dichotomically stable for real values of λ (fig. 2).

Two families of schemes have the characteristic polynomial (10). Dichotomically stable second derivative schemes:

$$Y_{i+1}=Y_i+h\,\Sigma_{j=0}^{k}\beta_j\dot{Y}_{i-j+1}+h^2\gamma(\ddot{Y}_{i+1}-\ddot{Y}_i) \tag{11}$$

have local truncation error $O(h^{p+1})$ as $h\to 0$. Hybrid Implicit Dichotomic schemes consist of two linked formulae:

$$Y_{i+\theta}=\Sigma_{j=0}^{k}a_jY_{i-j+1}+h\alpha(\dot{Y}_{i+1}-\dot{Y}_i) \tag{12}$$

$$Y_{i+1}=Y_i+h\,\Sigma_{j=0}^{k}b_j\dot{Y}_{i-j+1}+h\beta\dot{Y}_{i+\theta} \tag{13}$$

which must be solved simultaneously. The coefficients in (12) may be chosen to give an interpolation with truncation error $O(h^p)$ as $h\to 0$. The quadrature rule (13) has truncation error $O(h^{p+1})$ and the coefficients satisfy $\beta_j=\beta a_j+b_j$ $(j=0,1,\dots,k)$ and $\gamma=\beta\alpha$. The parameter θ may take any non-integer value, and in particular it may be chosen to make $b_k=0$.

For the case $k=2$, it happens that $\beta_2=0$, and the resulting formulae are one-step A-stable schemes, the symmetric second derivative method:

$$Y_{i+1}=Y_i+\frac{1}{2}h(\dot{Y}_{i+1}+\dot{Y}_i)-\frac{1}{12}h^2(\ddot{Y}_{i+1}-\ddot{Y}_i) \tag{14}$$

and the implicit Runge-Kutta (Lobatto collocation) scheme:

$$\begin{aligned} Y_{i+1/2} &= \frac{1}{2}(Y_{i+1}+Y_i)-\frac{1}{8}h(\dot{Y}_{i+1}-\dot{Y}_i)\\ Y_{i+1} &= Y_i+h\left[\frac{1}{6}\dot{Y}_i+\frac{2}{3}\dot{Y}_{i+1/2}+\frac{1}{6}\dot{Y}_{i+1}\right] \end{aligned} \tag{15}$$

which has made several appearances in the literature, e.g. [12], [10].

5. Matched Embedded Error Indicator

To implement this scheme in a variable step size mode, another method is needed, with order at least 3, so that the difference, which

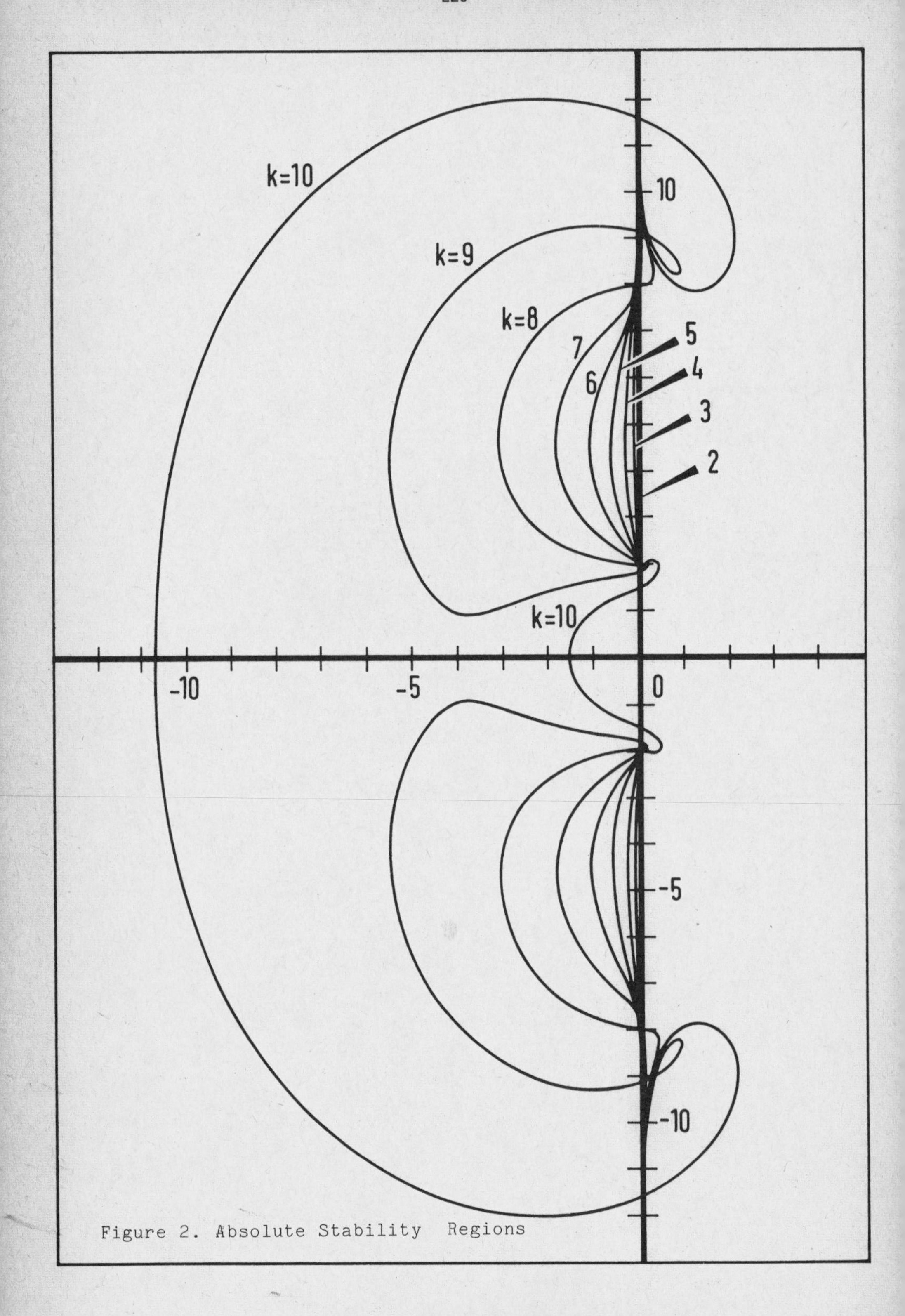

Figure 2. Absolute Stability Regions

is $O(h^4)$ as $h \to 0$, may be used as an error indicator to control the step size.

An explicit Runge-Kutta predictor might be considered, but it is important to consider the need to approximate, outside sharp layers, the smooth particular solution to $\dot{y}=\lambda y+f(t)$ using a step size $h \gg 1/\lambda$. Applying (15) to this test problem gives:

$$y_{i+1}=y_i \frac{1+\frac{1}{2} h\lambda+\frac{1}{12} h^2\lambda^2}{1-\frac{1}{2} h\lambda+\frac{1}{12} h^2\lambda^2}$$

$$+ h\frac{\frac{1}{6} f_i+\frac{2}{3} f_{i+1/2}+\frac{1}{6} f_{i+1}- \frac{1}{12} h\lambda(f_{i+1}-f_i)}{1 -\frac{1}{2} h\lambda+\frac{1}{12} h^2\lambda^2}$$

$$\sim \quad y_i + \frac{f_i}{\lambda} - \frac{f_{i+1}}{\lambda} \quad \text{as} \quad |h\lambda| \to \infty \quad , \qquad (16)$$

and thus generates a solution almost parallel to the smooth particular solution. An explicit Runge-Kutta predictor would generate:

$$y^P_{i+1} = y_i \times \text{polynomial in } h\lambda$$

$$+ hf \times \text{polynomial in } h\lambda$$

$$\to \infty \quad \text{as} \quad |h\lambda| \to \infty \quad ,$$

and thus give a large error indicator, and force a reduction of step size until $h=O(1/\lambda)$.

At fixed step size, the explicit 4 step extrapolator:

$$Y^P_{i+1} = 4Y_i - 6Y_{i-1} + 4Y_{i-2} - Y_{i-3} \qquad (17)$$

using past data generated by (15), gives:

$$y^P_{i+1} \sim y_i + \frac{f_i}{\lambda} - \frac{4f_i - 6f_{i-1} + 4f_{i-2} - f_{i-3}}{\lambda} \quad \text{as } |h\lambda| \to \infty$$

$$= y_i + \frac{f_i}{\lambda} - \frac{f_{i+1}}{\lambda} + \frac{h^4 f^{(iv)}_{i-1}}{\lambda} + O(\frac{h^5}{\lambda}) \quad \text{as } h \to 0 \quad , \qquad (18)$$

which is a third order approximation to a curve almost parallel to the smooth particular solution. It is thus a suitable third order predictor, and by interpolation theory, the corresponding coefficients are easily found for the variable step size case.

As implied by the name "predictor", (17) may be used not only for "estimating" the error, but also to obtain an initial value of y_{i+1} for iterative solution of (15). However, on the first three steps of any particular solution, there will be insufficient past data for the use of (17). It is then necessary to use a lower order predictor of the same type, although no error indicator will be available until four steps are completed. If the fourth step is rejected then all four steps must be rejected, but thereafter the error indicator can be evaluated at every step, and the step rejected, or the step size reduced as appropriate. The step size would only be increased if the norm of the error indicator is less than half the tolerance specified, for five consecutive steps.

6. Corrector Iteration Techniques

For non-linear problems it is necessary to solve equations (15) iteratively, and if the Jacobian matrix $\partial G/\partial Y$ has large eigenvalues, this must be done by some modified Newton iteration in order to preserve the stability properties. Using (1) to eliminate $Y_{i+1/2}$ from (15), the equation for Y_{i+1} is:

$$Y_{i+1}-Y_i-h[\frac{1}{6}\dot{Y}_{i+1}+\frac{2}{3}G\{\frac{1}{2}(Y_{i+1}+Y_i)-\frac{1}{8}h(\dot{Y}_{i+1}-\dot{Y}_i)\}+\frac{1}{6}\dot{Y}_i] = 0 \qquad (19)$$

and the iteration matrix must be $I-\frac{1}{2}hJ+\frac{1}{2}h^2J^2$ where J is a recent value of $\partial G/\partial Y$, which must be reevaluated when convergence becomes slow. For problems with a constant Jacobian matrix, convergence is then obtained in one iteration.

The rate of convergence may be estimated at every iteration from consecutive values of the residual of (19). While convergence is likely within three more iterations, the process may continue. If it becomes slow even with an updated value of J, then the step size must be reduced to a level at which convergence may be expected in one iteration.

The iteration may be considered to have converged when both the residual and the Newton step have a norm less than TOL/8, where TOL is the tolerance for the local error indicator. To avoid an extra evaluation of the function G, this last Newton step need not be applied.

With the strategy outlined above, the step size h will usually change more frequently than the matrix J is evaluated. It is thus desirable, unless J has some special sparsity structure which would be destroyed, to perform a similarity transformation upon it to some simpler form, as soon as it is evaluated. An appropriate transformation [9] uses a lower triangular matrix L to transform J to upper Hessenberg form U:

$$J - LUL^{-1} \tag{20}$$

with appropriate pivoting so that elements of L, L^{-1} do not exceed unity. Then

$$J^2-6J/h+12I/h^2=L[U^2-6U/h+12I/h^2]L^{-1} \quad . \tag{21}$$

The expression in square brackets must still be factorized for each new value of h. However, approximation by a perfect square, as proposed in [16], would be unsuitable, as the resulting iteration would not converge quadratically or indeed at all in a region around the artificially introduced real pole in $R_1(z)$. There appear to be two possible factorizations.

Firstly, a straightforward triangular factorization:

$$U^2-6U/h + 12I/h^2 = L_2R \tag{22}$$

which is particularly cheap, since the lower triangular factor L_2 has only two non-zero elements below the diagonal in each column. The number of multiplications required to form U, by (20), and U^2 is $n^3+0(n^2)$ as $n\to\infty$. Formation of 6U/h and factorization by (22) costs $\frac{3}{2}n^2+0(n)$ multiplications per step size h. Back substitution with the factorized matrix, to obtain the Newton step from the residual vector, costs another $\frac{3}{2}n^2+0(n)$ multiplications per iteration. This appears to be the cheapest possible factorization if the total number of iterations is more than n/9 for each evaluation of J. However, if hJ (and hU) have very large eigenvalues, U^2 will dominate the matrix in (22), and its condition number will be too large for the iteration to converge.

The other factorization is into complex factors, one of which is further reduced to triangular factors:

$$\begin{aligned} U^2-6U/h+12I/h^2 &= [U-(3+i\sqrt{3})I/h][U-(3-i\sqrt{3})I/h] \\ &= L_1R[U-(3-i\sqrt{3})I/h] \end{aligned} \tag{23}$$

where the complex valued lower triangular matrix L_1 has only one non-zero element below the diagonal in each column. The number of multiplications required to form U by (20) is $\frac{5}{6}n^3+0(n^2)$. Complex factorization of $[U-(3+i\sqrt{3})I/h]$ costs $2n^2+0(n)$ real multiplications per step size h (or $n^2/2$ complex multiplications). Back substitution with the second complex factor costs only n real multiplications [9], and so the total back substitution process costs $3n^2+O(n)$ real multiplications per iteration. It is this factorization which has been implemented in the tests which are now described.

7. Test Problem and Preliminary Results

Test cases have been run, solving initial value problems of the type:

$$\dot{Y} = JY + F(t) \tag{24}$$

with initial values close to the smooth particular solution. The smooth solution is followed for a considerable length of time before the step size is drastically reduced, and the fast growing modes begin to be accurately represented.

The matrix J has been taken to be

$$J = \begin{bmatrix} 9 & -12 & -5 & 0 \\ -12 & 9 & 0 & -5 \\ 15 & 0 & 9 & -12 \\ 0 & 15 & -12 & -9 \end{bmatrix} \tag{25}$$

so that the complementary modes have growth like $\exp\{(\pm 15 \pm 5i\sqrt{3})t\}$. F(t) has been chosen so that $Y=\{0,0, \exp(20t/\mu), 2\exp(-20t/\mu)\}$ is an exact solution, and the system integrated with initial values on this solution, and at two different deviations from it. Four values of the "stiffness" ratio μ, and three values of TOL have also been used, the latter to show the relation between the initial perturbations and the local error tolerance.

The results in table 1 show the number of steps taken, the final value of t reached, the range of step sizes h used, and the global error observed at that point. When the integration was stopped through a dramatic reduction in step size owing to blow up of the unstable modes, this has been noted, but it only occurs when excessive precision is requested, or the initial perturbation exceeds the tolerance. In other cases, the global error may be explained as the sum of the initial perturbation, virtually unchanged, and the global truncation error, independent of the perturbation. The results demonstrate the feasibility of forward integration with large step sizes corresponding to the smooth particular solution, and not greatly affected by the presence of strongly unstable modes, at least until μ is so large that the conditioning of hJ (and hU) becomes a problem.

Acknowledgement

This work was partly supported by the British Science and Engineering Research Council (SERC) during a visit to Imperial College, London, and by the Netherlands Organization for the Advancement of Pure Research (ZWO).

TABLE 1

Results for test problem:

No. of steps.
final t.
range of h.
global error.

Exact Initial Values at t=0

TOL	10^{-4}	10^{-7}	10^{-10}
$\mu=10^4$	14 5×10^2 [33.2,40] 6×10^{-9}	81 5×10^2 [4,6.85] 4×10^{-9}	9 5.0 [0.4,0.759] 2×10^{-11} Collapse $h\approx10^{-2}$
$\mu=10^5$	17 5×10^3 $[2\times10^2,3.79\times10^2]$ 6×10^{-10}	84 5×10^3 [20,68.4] 2×10^{-11}	63 2.9×10^3 [2,12.2] 2×10^{-9} Collapse $h\approx10^{-3}$
$\mu=10^6$	20 5×10^4 $[1\times10^3,3.74\times10^3]$ 5×10^{-11}	121 5×10^4 $[1\times10^2,6.7\times10^2]$ 1×10^{-12}	208 2.6×10^3 [10,23.1] 5×10^{-10} Collapse $h\approx10^{-2}$
$\mu=10^7$	45 5×10^5 $[5\times10^3,1.82\times10^4]$ 7×10^{-13}	521 5×10^5 $[3.29\times10^2,1.98\times10^3]$ 8×10^{-15}	280 3.4×10^3 [7.28,31.6] 2×10^{-8} Collapse $h\approx10^{-4}$

TABLE 1 (continued a)

Initial Values Perturbed by approximately 10^{-9}

TOL	10^{-4}	10^{-7}	10^{-10}
$\mu=10^4$	14 5×10^2 [33.2,40] 1×10^{-9}	81 5×10^2 [4,6.85] 2×10^{-7}	866 1.25 $[3.21\times10^{-4},2.5\times10^{-2}]$ 9×10^{-3}
$\mu=10^5$	17 5×10^3 $[2\times10^2,3.79\times10^2]$ 6×10^{-10}	84 5×10^3 [20,68.4] 1×10^{-9}	0 0.0 Collapse $h\approx10^{-2}$
$\mu=10^6$	20 5×10^4 $[1\times10^3,3.74\times10^3]$ 1×10^{-9}	133 5×10^4 $[1\times10^2,6.79\times10^2]$ 1×10^{-9}	6 67.4 [10,17.4] Collapse $h\approx10^{-2}$
$\mu=10^7$	45 5×10^5 $[5\times10^3,1.81\times10^4]$ 1×10^{-9}	1332 5×10^5 $[2.75\times10^2,7.58\times10^2]$ 1×10^{-9}	16 2×10^2 [11.9,15.5] 5×10^{-9} Collapse $h\approx10^{-3}$

TABLE 1 (continued b)

Initial Values Perturbed by 5×10^{-6}

TOL	10^{-4}	10^{-7}	10^{-10}
$\mu=10^4$	14 5×10^2 [33.2,40] 5×10^{-6}	5 20 [4,4] Collapse $h\approx10^{-3}$	0 0.0 Collapse $h\approx10^{-3}$
$\mu=10^5$	17 5×10^3 [2×10^2,3.79×10^2] 5×10^{-6}	240 4.8×10^3 [20,20] Collapse $h\approx10^{-5}$	0 0.0 Collapse $h\approx10^{-3}$
$\mu=10^6$	20 5×10^4 [1×10^3,3.74×10^3] 5×10^{-6}		
$\mu=10^7$	42 5×10^4 [5×10^3,2.42×10^4] 5×10^{-6}		

8. References

[1] U. Ascher, J. Christiansen, R.D. Russell, Math. Comp., Vol. 33, pp. 659-679 (1978).

[2] W.A. Coppel, Lecture Notes in Mathematics, Vol. 629, Springer (1978).

[3] R. England, Computer J., Vol. 12, pp. 166-170 (1969).

[4] R. England, UKAEA Culham Laboratory Report PDN 3/73 (1976).

[5] R. England, in Numerical Treatment of Inverse Problems for Differential and Integral Equations (eds. P. Deuflhard, E. Hairer), pp. 122-136, Series Progress in Scientific Computing, Vol. 2, Birkhäuser (1983).

[6] R. England, R.M.M. Mattheij, K.U. Nijmegen Dept. of Maths. Report 8356 (1983).

[7] R. England, R.M.M. Mattheij, K.U. Nijmegen Dept. of Maths. Report 8439 (1984).

[8] R. England, R.M.M. Mattheij, to be presented in Workshop on Numerical Analysis and Its Applications, IVth Mathematics Colloquium, Centro de Investigación y de Estudios Avanzados del Instituto Politécnico Nacional, Taxco, México (1985).

[9] W.H. Enright, ACM Trans. Math. Software, Vol. 4, pp. 127-136 (1978).

[10] J.P. Hennart, R. England, in Working Papers 1979 SIGNUM Meeting on Numerical O.D.E.'s (ed. R.D. Skeel), pp. 33.1-33.4, U. of Illinois at Urbana-Champaign Dept. of Computer Science. Report 963 (1979).

[11] F.R. de Hoog, R.M.M. Mattheij, K.U. Nijmegen Dept. of Maths. Report 8355 (1983).

[12] K.S. Kunz: Numerical Analysis, 1st. Edition, p. 206, McGraw-Hill (1957).

[13] M. Lentini, V. Pereyra, SIAM J. Numer. Anal., Vol. 14, pp. 91-111 (1977).

[14] R.M.M. Mattheij, SIAM J. Numer. Anal., Vol. 19, pp. 963-978 (1982).

[15] N.K. Nichols, R. England, J. Comput. Phys., Vol. 46, pp. 369-389 (1982).

[16] R.D. Skeel, A.K. Kong, ACM Trans. Math. Software, Vol. 3, pp. 326-345 (1977).

Vol. 1062: J. Jost, Harmonic Maps Between Surfaces. X, 133 pages. 1984.

Vol. 1063: Orienting Polymers. Proceedings, 1983. Edited by J.L. Ericksen. VII, 166 pages. 1984.

Vol. 1064: Probability Measures on Groups VII. Proceedings, 1983. Edited by H. Heyer. X, 588 pages. 1984.

Vol. 1065: A. Cuyt, Padé Approximants for Operators: Theory and Applications. IX, 138 pages. 1984.

Vol. 1066: Numerical Analysis. Proceedings, 1983. Edited by D.F. Griffiths. XI, 275 pages. 1984.

Vol. 1067: Yasuo Okuyama, Absolute Summability of Fourier Series and Orthogonal Series. VI, 118 pages. 1984.

Vol. 1068: Number Theory, Noordwijkerhout 1983. Proceedings. Edited by H. Jager. V, 296 pages. 1984.

Vol. 1069: M. Kreck, Bordism of Diffeomorphisms and Related Topics. III, 144 pages. 1984.

Vol. 1070: Interpolation Spaces and Allied Topics in Analysis. Proceedings, 1983. Edited by M. Cwikel and J. Peetre. III, 239 pages. 1984.

Vol. 1071: Padé Approximation and its Applications, Bad Honnef 1983. Prodeedings. Edited by H. Werner and H.J. Bünger. VI, 264 pages. 1984.

Vol. 1072: F. Rothe, Global Solutions of Reaction-Diffusion Systems. V, 216 pages. 1984.

Vol. 1073: Graph Theory, Singapore 1983. Proceedings. Edited by K.M. Koh and H.P. Yap. XIII, 335 pages. 1984.

Vol. 1074: E.W. Stredulinsky, Weighted Inequalities and Degenerate Elliptic Partial Differential Equations. III, 143 pages. 1984.

Vol. 1075: H. Majima, Asymptotic Analysis for Integrable Connections with Irregular Singular Points. IX, 159 pages. 1984.

Vol. 1076: Infinite-Dimensional Systems. Proceedings, 1983. Edited by F. Kappel and W. Schappacher. VII, 278 pages. 1984.

Vol. 1077: Lie Group Representations III. Proceedings, 1982–1983. Edited by R. Herb, R. Johnson, R. Lipsman, J. Rosenberg. XI, 454 pages. 1984.

Vol. 1078: A.J.E.M. Janssen, P. van der Steen, Integration Theory. V, 224 pages. 1984.

Vol. 1079: W. Ruppert. Compact Semitopological Semigroups: An Intrinsic Theory. V, 260 pages. 1984

Vol. 1080: Probability Theory on Vector Spaces III. Proceedings, 1983. Edited by D. Szynal and A. Weron. V, 373 pages. 1984.

Vol. 1081: D. Benson, Modular Representation Theory: New Trends and Methods. XI, 231 pages. 1984.

Vol. 1082: C.-G. Schmidt, Arithmetik Abelscher Varietäten mit komplexer Multiplikation. X, 96 Seiten. 1984.

Vol. 1083: D. Bump, Automorphic Forms on GL (3,IR). XI, 184 pages. 1984.

Vol. 1084: D. Kletzing, Structure and Representations of Q-Groups. VI, 290 pages. 1984.

Vol. 1085: G.K. Immink, Asymptotics of Analytic Difference Equations. V, 134 pages. 1984.

Vol. 1086: Sensitivity of Functionals with Applications to Engineering Sciences. Proceedings, 1983. Edited by V. Komkov. V, 130 pages. 1984

Vol. 1087: W. Narkiewicz, Uniform Distribution of Sequences of Integers in Residue Classes. VIII, 125 pages. 1984.

Vol. 1088: A.V. Kakosyan, L.B. Klebanov, J.A. Melamed, Characterization of Distributions by the Method of Intensively Monotone Operators. X, 175 pages. 1984.

Vol. 1089: Measure Theory, Oberwolfach 1983. Proceedings. Edited by D. Kölzow and D. Maharam-Stone. XIII, 327 pages. 1984.

Vol. 1090: Differential Geometry of Submanifolds. Proceedings, 1984. Edited by K. Kenmotsu. VI, 132 pages. 1984.

Vol. 1091: Multifunctions and Integrands. Proceedings, 1983. Edited by G. Salinetti. V, 234 pages. 1984.

Vol. 1092: Complete Intersections. Seminar, 1983. Edited by S. Greco and R. Strano. VII, 299 pages. 1984.

Vol. 1093: A. Prestel, Lectures on Formally Real Fields. XI, 125 pages. 1984.

Vol. 1094: Analyse Complexe. Proceedings, 1983. Edité par E. Amar, R. Gay et Nguyen Thanh Van. IX, 184 pages. 1984.

Vol. 1095: Stochastic Analysis and Applications. Proceedings, 1983. Edited by A. Truman and D. Williams. V, 199 pages. 1984.

Vol. 1096: Théorie du Potentiel. Proceedings, 1983. Edité par G. Mokobodzki et D. Pinchon. IX, 601 pages. 1984.

Vol. 1097: R.M. Dudley, H. Kunita, F. Ledrappier, École d'Éte de Probabilités de Saint-Flour XII – 1982. Edité par P.L. Hennequin. X, 396 pages. 1984.

Vol. 1098: Groups – Korea 1983. Proceedings. Edited by A.C. Kim and B.H. Neumann. VII, 183 pages. 1984.

Vol. 1099: C.M. Ringel, Tame Algebras and Integral Quadratic Forms. XIII, 376 pages. 1984.

Vol. 1100: V. Ivrii, Precise Spectral Asymptotics for Elliptic Operators Acting in Fiberings over Manifolds with Boundary. V, 237 pages. 1984.

Vol. 1101: V. Cossart, J. Giraud, U. Orbanz, Resolution of Surface Singularities. Seminar. VII, 132 pages. 1984.

Vol. 1102: A. Verona, Stratified Mappings – Structure and Triangulability. IX, 160 pages. 1984.

Vol. 1103: Models and Sets. Proceedings, Logic Colloquium, 1983, Part I. Edited by G.H. Müller and M.M. Richter. VIII, 484 pages. 1984.

Vol. 1104: Computation and Proof Theory. Proceedings, Logic Colloquium, 1983, Part II. Edited by M.M. Richter, E. Börger, W. Oberschelp, B. Schinzel and W. Thomas. VIII, 475 pages. 1984.

Vol. 1105: Rational Approximation and Interpolation. Proceedings, 1983. Edited by P.R. Graves-Morris, E.B. Saff and R.S. Varga. XII, 528 pages. 1984.

Vol. 1106: C.T. Chong, Techniques of Admissible Recursion Theory. IX, 214 pages. 1984.

Vol. 1107: Nonlinear Analysis and Optimization. Proceedings, 1982. Edited by C. Vinti. V, 224 pages. 1984.

Vol. 1108: Global Analysis – Studies and Applications I. Edited by Yu.G. Borisovich and Yu.E. Gliklikh. V, 301 pages. 1984.

Vol. 1109: Stochastic Aspects of Classical and Quantum Systems. Proceedings, 1983. Edited by S. Albeverio, P. Combe and M. Sirugue-Collin. IX, 227 pages. 1985.

Vol. 1110: R. Jajte, Strong Limit Theorems in Non-Commutative Probability. VI, 152 pages. 1985.

Vol. 1111: Arbeitstagung Bonn 1984. Proceedings. Edited by F. Hirzebruch, J. Schwermer and S. Suter. V, 481 pages. 1985.

Vol. 1112: Products of Conjugacy Classes in Groups. Edited by Z. Arad and M. Herzog. V, 244 pages. 1985.

Vol. 1113: P. Antosik, C. Swartz, Matrix Methods in Analysis. IV, 114 pages. 1985.

Vol. 1114: Zahlentheoretische Analysis. Seminar. Herausgegeben von E. Hlawka. V, 157 Seiten. 1985.

Vol. 1115: J. Moulin Ollagnier, Ergodic Theory and Statistical Mechanics. VI, 147 pages. 1985.

Vol. 1116: S. Stolz, Hochzusammenhängende Mannigfaltigkeiten und ihre Ränder. XXIII, 134 Seiten. 1985.

Vol. 1117: D.J. Aldous, J.A. Ibragimov, J. Jacod, Ecole d'Été de Probabilités de Saint-Flour XIII – 1983. Édité par P.L. Hennequin. IX, 409 pages. 1985.

Vol. 1118: Grossissements de filtrations: exemples et applications. Seminaire, 1982/83. Edité par Th. Jeulin et M. Yor. V, 315 pages. 1985.

Vol. 1119: Recent Mathematical Methods in Dynamic Programming. Proceedings, 1984. Edited by I. Capuzzo Dolcetta, W.H. Fleming and T. Zolezzi. VI, 202 pages. 1985.

Vol. 1120: K. Jarosz, Perturbations of Banach Algebras. V, 118 pages. 1985.

Vol. 1121: Singularities and Constructive Methods for Their Treatment. Proceedings, 1983. Edited by P. Grisvard, W. Wendland and J.R. Whiteman. IX, 346 pages. 1985.

Vol. 1122: Number Theory. Proceedings, 1984. Edited by K. Alladi. VII, 217 pages. 1985.

Vol. 1123: Séminaire de Probabilités XIX 1983/84. Proceedings. Edité par J. Azéma et M. Yor. IV, 504 pages. 1985.

Vol. 1124: Algebraic Geometry, Sitges (Barcelona) 1983. Proceedings. Edited by E. Casas-Alvero, G.E. Welters and S. Xambó-Descamps. XI, 416 pages. 1985.

Vol. 1125: Dynamical Systems and Bifurcations. Proceedings, 1984. Edited by B.L.J. Braaksma, H.W. Broer and F. Takens. V, 129 pages. 1985.

Vol. 1126: Algebraic and Geometric Topology. Proceedings, 1983. Edited by A. Ranicki, N. Levitt and F. Quinn. V, 423 pages. 1985.

Vol. 1127: Numerical Methods in Fluid Dynamics. Seminar. Edited by F. Brezzi, VII, 333 pages. 1985.

Vol. 1128: J. Elschner, Singular Ordinary Differential Operators and Pseudodifferential Equations. 200 pages. 1985.

Vol. 1129: Numerical Analysis, Lancaster 1984. Proceedings. Edited by P.R. Turner. XIV, 179 pages. 1985.

Vol. 1130: Methods in Mathematical Logic. Proceedings, 1983. Edited by C.A. Di Prisco. VII, 407 pages. 1985.

Vol. 1131: K. Sundaresan, S. Swaminathan, Geometry and Nonlinear Analysis in Banach Spaces. III, 116 pages. 1985.

Vol. 1132: Operator Algebras and their Connections with Topology and Ergodic Theory. Proceedings, 1983. Edited by H. Araki, C.C. Moore, Ş. Strătilă and C. Voiculescu. VI, 594 pages. 1985.

Vol. 1133: K.C. Kiwiel, Methods of Descent for Nondifferentiable Optimization. VI, 362 pages. 1985.

Vol. 1134: G.P. Galdi, S. Rionero, Weighted Energy Methods in Fluid Dynamics and Elasticity. VII, 126 pages. 1985.

Vol. 1135: Number Theory, New York 1983–84. Seminar. Edited by D.V. Chudnovsky, G.V. Chudnovsky, H. Cohn and M.B. Nathanson. V, 283 pages. 1985.

Vol. 1136: Quantum Probability and Applications II. Proceedings, 1984. Edited by L. Accardi and W. von Waldenfels. VI, 534 pages. 1985.

Vol. 1137: Xiao G., Surfaces fibrées en courbes de genre deux. IX, 103 pages. 1985.

Vol. 1138: A. Ocneanu, Actions of Discrete Amenable Groups on von Neumann Algebras. V, 115 pages. 1985.

Vol. 1139: Differential Geometric Methods in Mathematical Physics. Proceedings, 1983. Edited by H. D. Doebner and J. D. Hennig. VI, 337 pages. 1985.

Vol. 1140: S. Donkin, Rational Representations of Algebraic Groups. VII, 254 pages. 1985.

Vol. 1141: Recursion Theory Week. Proceedings, 1984. Edited by H.-D. Ebbinghaus, G.H. Müller and G.E. Sacks. IX, 418 pages. 1985.

Vol. 1142: Orders and their Applications. Proceedings, 1984. Edited by I. Reiner and K. W. Roggenkamp. X, 306 pages. 1985.

Vol. 1143: A. Krieg, Modular Forms on Half-Spaces of Quaternions. XIII, 203 pages. 1985.

Vol. 1144: Knot Theory and Manifolds. Proceedings, 1983. Edited by D. Rolfsen. V, 163 pages. 1985.

Vol. 1145: G. Winkler, Choquet Order and Simplices. VI, 1985.

Vol. 1146: Séminaire d'Algèbre Paul Dubreil et Marie-Pa Proceedings, 1983–1984. Edité par M.-P. Malliavin. IV, 1985.

Vol. 1147: M. Wschebor, Surfaces Aléatoires. VII, 111 pag

Vol. 1148: Mark A. Kon, Probability Distributions in Quantu Mechanics. V, 121 pages. 1985.

Vol. 1149: Universal Algebra and Lattice Theory. Procee Edited by S. D. Comer. VI, 282 pages. 1985.

Vol. 1150: B. Kawohl, Rearrangements and Convexity of L PDE. V, 136 pages. 1985.

Vol 1151: Ordinary and Partial Differential Equations. P 1984. Edited by B.D. Sleeman and R.J. Jarvis. XIV, 357 pa

Vol. 1152: H. Widom, Asymptotic Expansions for Pseud Operators on Bounded Domains. V, 150 pages. 1985.

Vol. 1153: Probability in Banach Spaces V. Proceedings, 1 by A. Beck, R. Dudley, M. Hahn, J. Kuelbs and M. Marc pages. 1985.

Vol. 1154: D.S. Naidu, A.K. Rao, Singular Pertubation Discrete Control Systems. IX, 195 pages. 1985.

Vol. 1155: Stability Problems for Stochastic Models. Pr 1984. Edited by V.V. Kalashnikov and V.M. Zolotare pages. 1985.

Vol. 1156: Global Differential Geometry and Global Anal Proceedings, 1984. Edited by D. Ferus, R.B. Gardner, S. and U. Simon. V, 339 pages. 1985.

Vol. 1157: H. Levine, Classifying Immersions into $\mathbb{R}^4$ over St of 3-Manifolds into $\mathbb{R}^2$. V, 163 pages. 1985.

Vol. 1158: Stochastic Processes – Mathematics and Physic dings, 1984. Edited by S. Albeverio, Ph. Blanchard and L. 230 pages. 1986.

Vol. 1159: Schrödinger Operators, Como 1984. Seminar. E Graffi. VIII, 272 pages. 1986.

Vol. 1160: J.-C. van der Meer, The Hamiltonian Hopf Bifur 115 pages. 1985.

Vol. 1161: Harmonic Mappings and Minimal Immersions, M 1984. Seminar. Edited by E. Giusti. VII, 285 pages. 1985.

Vol. 1162: S.J.L. van Eijndhoven, J. de Graaf, Trajectory Generalized Functions and Unbounded Operators. IV, 27 1985.

Vol. 1163: Iteration Theory and its Functional Equations. Pro 1984. Edited by R. Liedl, L. Reich and Gy. Targonski. VIII, 2 1985.

Vol. 1164: M. Meschiari, J.H. Rawnsley, S. Salamon, Seminar "Luigi Bianchi" II – 1984. Edited by E. Vesentini. pages. 1985.

Vol. 1165: Seminar on Deformations. Proceedings, 1982/8 by J. Ławrynowicz. IX, 331 pages. 1985.

Vol. 1166: Banach Spaces. Proceedings, 1984. Edited by and E. Saab. VI, 199 pages. 1985.

Vol. 1167: Geometry and Topology. Proceedings, 1983–84. J. Alexander and J. Harer. VI, 292 pages. 1985.

Vol. 1168: S.S. Agaian, Hadamard Matrices and their Applica 227 pages. 1985.

Vol. 1169: W.A. Light, E.W. Cheney, Approximation Theory Product Spaces. VII, 157 pages. 1985.

Vol. 1170: B.S. Thomson, Real Functions. VII, 229 pages. 19

Vol. 1171: Polynômes Orthogonaux et Applications. Proc 1984. Edité par C. Brezinski, A. Draux, A.P. Magnus, P. Ma Ronveaux. XXXVII, 584 pages. 1985.

Vol. 1172: Algebraic Topology, Göttingen 1984. Proceeding by L. Smith. VI, 209 pages. 1985.